p-Adic Methods and Their Applications

p-Adic Methods and Their Applications

Edited by

Andrew J. Baker

and

Roger J. Plymen

Department of Mathematics
University of Manchester
Manchester, UK

OXFORD · NEW YORK · TOKYO
CLARENDON PRESS
1992

This book has been printed digitally in order to ensure its continuing availability

OXFORD
UNIVERSITY PRESS

Great Clarendon Street, Oxford OX2 6DP

Oxford University Press is a department of the University of Oxford.
It furthers the University's objective of excellence in research, scholarship,
and education by publishing worldwide in

Oxford New York

Auckland Bangkok Buenos Aires Cape Town Chennai
Dar es Salaam Delhi Hong Kong Istanbul Karachi Kolkata
Kuala Lumpur Madrid Melbourne Mexico City Mumbai Nairobi
São Paulo Shanghai Singapore Taipei Tokyo Toronto

with an associated company in Berlin

Oxford is a registered trade mark of Oxford University Press
in the UK and in certain other countries

Published in the United States
by Oxford University Press Inc., New York

Reprinted 2002

A catalogue record for this book is available from the British Library

Library of Congress Cataloging in Publication Data
(Data available)

ISBN 0-19-853594-5

PREFACE

The p-adic numbers and more generally local fields have become increasingly important in a wide range of mathematical disciplines, and are now seen as essential tools in many areas of mathematics, including number theory, algebraic geometry, group representation theory, the modern theory of automorphic forms and algebraic topology.

A number of texts have recently become available which provide good general introductions to p-adic numbers and p-adic analysis. However, there is at present a gap between such books and the sophisticated applications in the research literature. The aim of this book is to bridge this gulf by providing a collection of intermediate-level articles on various applications of p-adic techniques throughout mathematics. We hope that the articles in this book will be especially useful for graduate students beginning research in such areas and we have encouraged all our contributors to write at a suitable level.

The idea for producing such a volume was suggested by Oxford University Press in connection with a three day meeting '**$\boldsymbol{p}$-adic Methods and their Applications**' held at Manchester University in September 1989 and which received financial support from the London Mathematical Society. Some of these articles grew out of talks given at this conference, others were written by invitation especially for this volume. All contributions were refereed with a particular view to their suitability for inclusion in such a book.

We would like to thank all of our contributors and referees for their efforts on behalf of this book. We would also like to thank the London Mathematical Society for its support, the Mathematics Department of Manchester University, the Science and Engineering Research Council for funding of the first editor, and finally all of our typesetters and local TEXperts.

DEDICATION TO THE MEMORY OF K. MAHLER (1903–1990)

One of the great names in the study of p-adic numbers died in early 1990. It seems very appropriate that we should dedicate this volume to his memory, since apart from his special place in the history of the subject, he also spent a significant part of his career in Manchester (1933, 1934 and 1937–1962) and indeed carried out much of his p-adic work there.

CONVENTIONS AND NOTATIONS

We use the symbols $\mathbb{N}, \mathbb{Z}, \mathbb{Q}, \mathbb{R}, \mathbb{C}$ for the natural numbers, the integers, the rational numbers, the real numbers and the complex numbers respectively. We also denote by $\mathbb{Z}_p$ and $\mathbb{Q}_p$ the p-adic integers and numbers respectively. Finally, $\mathbb{F}_{p^n}$ denotes the Galois field of p^n elements.

In this book, we will use the term *local field* to mean a commutative field F, equipped with a discrete valuation with respect to which F is locally compact. Such a field is necessarily complete and its maximal compact subring has finite residue field. Of course, this excludes the real numbers $\mathbb{R}$ and the complex numbers $\mathbb{C}$. The most familiar examples of local fields are the field of p-adic numbers $\mathbb{Q}_p$ and its finite extensions, and the field $\mathbb{F}_q((x))$ of finite tailed Laurent series in x over a finite field $\mathbb{F}_q$ where $q = p^n$ for a prime p.

Unfortunately, the term *local field* does not have a fixed meaning in the literature. Some authors relax the condition on the residue field, replacing it by the condition that it be perfect (for example, allowing the residue field to be $\mathbb{F}_{p^\infty}$, the algebraic closure of the finite field $\mathbb{F}_p$, or *any* characteristic 0 field). We must also emphasize that in the context of adèles and adèle groups the term local field has in recent years come to include archimedean fields. In the latter context, the reader will encounter the terms *local archimedean field*, *local non-archimedean field*, *local group* and indeed the term *local questions* as contrasted with global questions.

The term $\mathfrak{p}$-*adic field* is frequently used to mean a finite extension of the field of p-adic numbers $\mathbb{Q}_p$. Frustration with the Gothic $\mathfrak{p}$ (in pre-TEX days!) led to the use of the generic term *p-adic* to denote anything relating to non-archimedean mathematics, and indeed we use it in this sense in the title of our book.

To avoid confusion arising from such ambiguities, we have adopted the above strict definition of local field throughout this book.

Contents

1.

THE GRAY CODE FUNCTION

FRANCIS CLARKE

Department of Mathematics and Computer Science,
University College of Swansea,
Swansea SA2 8PP,
Wales.
`<mafred@uk.ac.swan.pyr>`

Figure 1.

Introduction

The *Gray code* of a positive integer m can be defined in terms of its binary expansion as follows. Shift m to the right, discarding the last digit, and then add the result to m without carrying (an exclusive or). Thus if we write $\oplus$ for the exclusive or operation and $\div$ for integer division, so that $m \div 2$ is the right shift of m, then the Gray code of m is $g(m) = m \oplus (m \div 2)$.

For each k, the function g permutes the integers between 0 and $2^k - 1$. It has the property that the binary expansions of $g(m)$ and $g(m+1)$ always differ in precisely one digit. This is true cyclically, in the sense that it holds for $g(2^k - 1) = 2^{k-1}$ and $g(0) = 0$. This provides a Hamiltonian circuit on the k-dimensional cube and is the basis of the construction of the diagram in Figure 1.

In this example $k = 5$. For each of $2^k = 32$ sectors the pattern in sector m represents the binary expansion of $g(m)$. The shaded areas represent conducting regions on a rotating disc. A radial row of brushes aligned to make a connection with these regions will then produce all 32 possible combinations of open and closed circuits. As the disc rotates only one brush will make or break a connection at a time. Thus there can be no ambiguity when the brushes cross the boundary between two sectors. It was for this purpose that Frank Gray's 1953 patent was designed [2]. The same idea had, it seems, already occurred to the Belgian telegraph engineer J. Émile Baudot in the 1870s; see [4]. The Gray code is commonly used (in optical form) in *shaft encoders*, which are used to measure angular displacement.

Acknowledgements: I should like to thank Alan Evans, Claude Haigh, David Singmaster and Chris Woodcock for several useful conversations about the material in this paper.

1 The extension to the dyadic integers

The function g can be extended, using the same definition, to a function $g : \mathbb{Z}_2 \to \mathbb{Z}_2$ on the 2-adic integers. If $x \equiv y \bmod 2^{k+1}$ then $g(x) \equiv g(y) \bmod 2^k$ so that g is continuous.

Write $x = x_0x_1 \ldots = \sum_{i=0}^{\infty} x_i 2^i$ in $\mathbb{Z}_2$, with $x_i = 0$ or 1, then $g(x) = \sum_{i=0}^{\infty}(x_i \oplus x_{i+1})2^i$. If we think of the expansion $x = x_0x_1 \ldots$ in terms of alternating blocks of zeros and ones then the expansion of $g(x)$ has ones at the positions of the last digits of each block of x. Note that, written this way, $x \mapsto x \div 2$ is a shift to the *left*. For example, if

$$\begin{array}{rcl} x & = & 0110001111000001111100\ldots, \\ \text{then} \quad g(x) & = & 1010010001000010000010\ldots. \end{array}$$

Thus if x is even and

$$x_i = \begin{cases} 0, & \text{if } k_{2j} < i \leqslant k_{2j+1}, \\ 1, & \text{if } k_{2j+1} < i \leqslant k_{2j+2}, \end{cases}$$

where we set $k_0 = -1$, we have $g(x) = \sum_{j\geqslant 1} 2^{k_j}$.

It follows immediately that, restricted to the even 2-adic integers, the function $g : 2\mathbb{Z}_2 \to \mathbb{Z}_2$ is bijective, and hence is a homeomorphism. Note also the functional equation $g(-1-x) = g(x)$. This follows since $-1-x = x \oplus (-1)$, which is formed by interchanging all the zeros and ones in the expansion of x.

Proposition 1.1. *The Gray code function g is nowhere differentiable.*

Proof We show here that $g(x)$ is not differentiable at rational integer values. We omit the proof for arbitrary 2-adic integers which is only slightly more involved. By the functional equation it is sufficient to prove that g is not differentiable at the positive integers. If $2^{k-1} > m$ then

$$\frac{g(m+2^k) - g(m)}{2^k} = \frac{2^{k-1} + 2^k}{2^k} = \frac{3}{2},$$

while

$$\frac{g(m+2^k+2^{k+1}) - g(m)}{2^k + 2^{k+1}} = \frac{2^{k-1} + 2^{k+1}}{2^k + 2^{k+1}} = \frac{5}{6}.$$

It follows that $\big(g(m+h) - g(m)\big)/h$ has no limit as $h \to 0$ in $\mathbb{Z}_2$. ◊ ◊

2 The interpolation series

By Mahler's theorem [3], we may write

$$g(x) = \sum_{n \geqslant 0} a_n \binom{x}{n},$$

for integers

$$a_n = \sum_{k=0}^{n} (-1)^{n-k} \binom{n}{k} g(k),$$

such that $a_n \to 0$ in the 2-adic topology as $n \to \infty$. In fact Bojanic's proof of Mahler's theorem (see [1], or Section 11 of Chapter 10 of [3]) shows that $\nu_2(a_n) \geqslant s$ if $n \geqslant s2^{s+1}$.

In conclusion, we present numerical evidence for some interesting regularity amongst these interpolation coefficients. We have not made much progress in explaining the phenomena which we illustrate, or even in formulating appropriate conjectures. Nevertheless the sequence of coefficients does exhibit a beautiful structure. We suggest that the Gray code function provides a useful example in the attempt to relate properties of p-adic functions to properties of their interpolation coefficients. The first few values are given in Table 1.

Table 1. The first interpolation coefficients of g

n	0	1	2	3	4	5	6	7	8	9	10	11
a_n	0	1	1	−4	12	−28	52	−80	112	−176	376	−976

It appears that, for $n > 1$, the sign of a_n alternates. But this is misleading; the pattern is more complex. For example, a_{43} is positive. After a_{43} the signs alternate again until a_{52} which is positive and the alternation of signs reverts to the original phase. Our computations suggest that $(-1)^n a_n$ changes sign infinitely often.

We now consider the rate of growth of $|a_n|$. In Figure 2 we plot the graph of $\log_2 |a_n|$ against n for n up to 300. This shows that $|a_n|$ grows very like 2^n. In view of the blocks of terms of alternating sign it is natural to consider the ratio $a_n/(-2)^n$.

In Figure 3 we plot the graph of $a_n/(-2)^n$ against n for $n = 0$ to $n = 2047$. The graph contains many intriguing patterns. In particular the regular spacing of the cumulatively largest maxima and the parabolic envelope in which they appear to lie are confirmed by Table 2. In this table we list these maxima. It can be seen that they occur at values of n close to the powers of 2. In the last column we give the values of $A\sqrt{n+B}+C$, where, to six decimal places, $A = 0.199\,220$, $B = 2.448\,112$ and $C =$

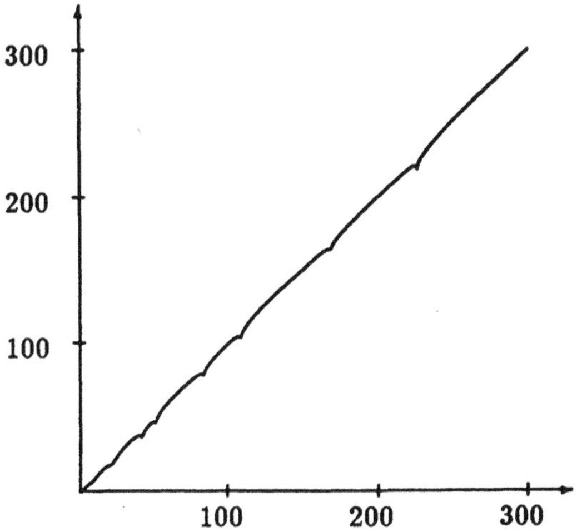

Figure 2. $\log_2 |a_n|$, for $n = 0$ to 300

Table 2. The local maxima of $a_n/(-2)^n$

n	$a_n/(-2)^n$	$A\sqrt{n+B}+C$
15	0.8379	0.8437
32	1.1788	1.1808
65	1.6535	1.6477
129	2.2959	2.2956
257	3.2208	3.2205
513	4.5331	4.5345
1025	6.3977	6.3973

0.011 545. This least-squares fit is very close, suggesting that the parabolic shape is more than coincidental.

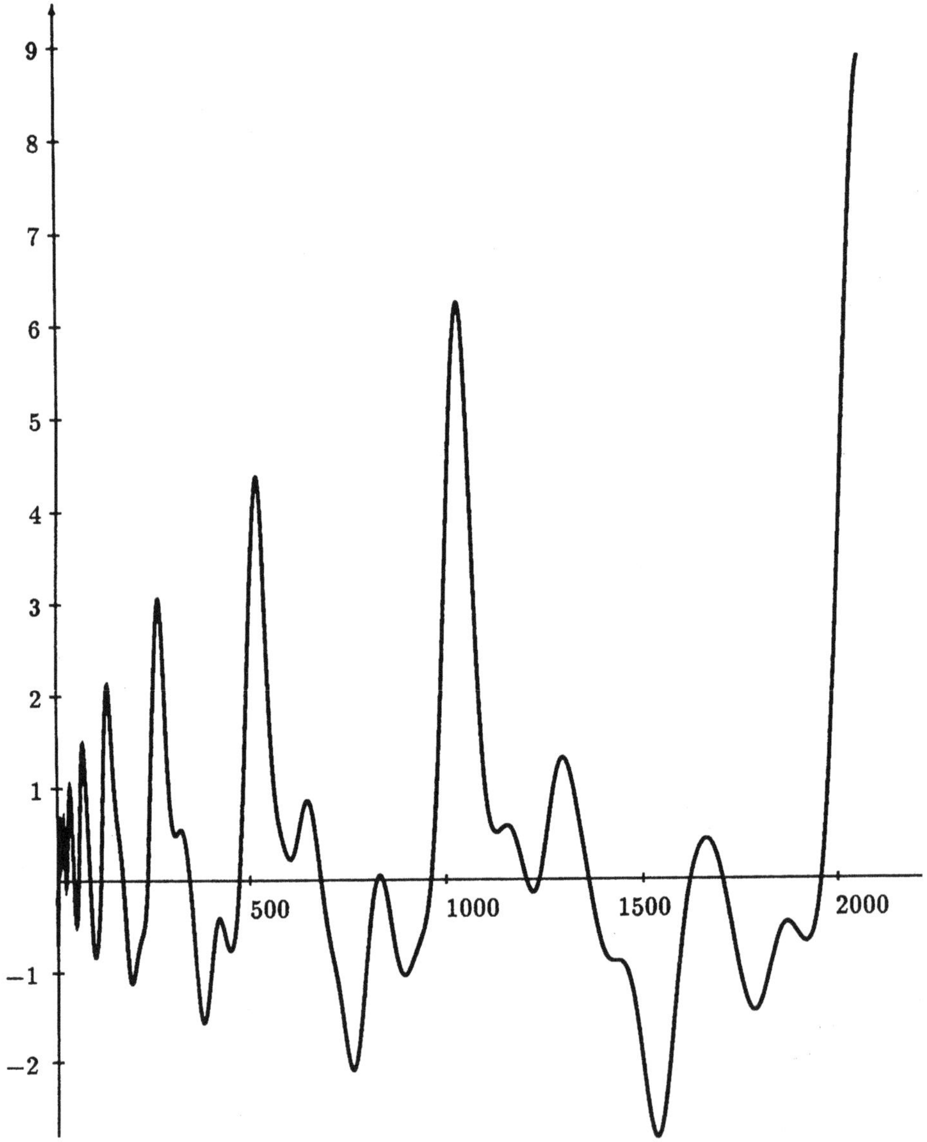

Figure 3. $a_n/(-2)^n$, for $n = 0$ to 2047

References

1. R. Bojanic, 'A simple proof of Mahler's theorem on the approximation of functions of a p-adic variable by polynomials', J. Number Theory **6** (1974), 412–415.

2. F. Gray, Pulse code communication, U. S. Patent 2.632.058 (1953).

3. K. Mahler, 'p-adic numbers and their functions', (2nd edn) Cambridge University Press (1981).

4. D. Singmaster, 'Binary Coding', (Topic area 11: Mathematical Games and Recreations) in Proceedings of Sixth International Cong. on Math. Education (Ed. Ann and Keith Hirst). ICME Secretariat and János Bolyai Math. Soc., Budapest (1988) 361–364.

2.

APPLICATIONS OF p-ADIC METHODS TO GROUP THEORY

MARCUS P. F. DU SAUTOY

All Soul's College,
Oxford OX1 4AL,
England.
<mdus@uk.ac.ox.vax>

Introduction

In the late nineteenth century, Sophus Lie studied 'transformation groups' in an attempt to understand various geometries from more group-theoretic point of view. His pioneering work forged what has been known for nearly a century as the theory of Lie groups, i.e. topological groups with the underlying structure of a real manifold with respect to which the group operations are analytic. The importance of the theory of Lie groups can be judged by the diversity of its applications in topology, differential geometry and arithmetic.

There have also been a host of parallel theories where the underlying real manifold is replaced by such structures as algebraic varieties, schemes and p-adic manifolds. It is the last of these which lies at the heart of this chapter. Such groups were encountered for the first time in 1907 in the writings of Hensel [17] on p-adic analytic functions. By 1936 they had appeared in work of Weil and Lutz on p-adic elliptic curves (see [50] and [29]).

However, it was the publication of Lazard's seminal paper 'Groupes analytiques p-adiques' [21] in 1965 which crystallized the theory of compact p-adic analytic groups. He showed that the existence of a p-adic analytic structure on a compact topological group is closely related to that of certain filtrations, concluding that the analytic nature of these groups is dictated by its algebraic behaviour. It is this algebraic approach which we shall adopt as our starting point. We then show among other things how the filtrations studied by Lazard are used to produce a p-adic analytic manifold and a natural Lie algebra associated with these groups. In this fashion the reader has to master little analytic machinery to gain access to the riches available in the theory of p-adic analytic groups.

The *raison d'être* for this chapter comes in the second section. Here, in contrast to Lie's project of studying geometry from a group-theoretic viewpoint, we present a survey of results in the theory of abstract groups which have been proved using p-adic analytic groups. The possibility for such an application is a construction detailed in the first section which provides a passage from abstract groups to the realm of p-adic analytic groups. This passage is then exploited to deduce a rich array of results in the theory of abstract groups ranging from the Congruence Subgroup Problem, to combinatorial group theory, and from a criterion for a group to be linear to rationality results for Poincaré series associated with groups. We hope that the philosophy proposed in this chapter will highlight a growing corner of group theory. The reader who finds all this icing unsatisfying is encouraged to peruse [7] which fleshes out much of this chapter, together with two excellent survey articles [31] and [45].

1 Analytic pro-p groups

A pro-p group is an inverse limit of finite p-groups. We begin by explaining this definition in some detail. There is a dichotomy in the theory of pro-p groups according to whether the pro-p group has an underlying analytic structure or not. We indicate a group-theoretic criterion which enables us to recognize a pro-p group with such a structure; this is Lazard's answer to the p-adic version of Hilbert's fifth problem.

1.1 Inverse limits

A *directed set* is a non-empty ordered set $(I, \leqslant)$ with the property that to every i_1, i_2 in I, there exists i in I such that $i_1 \leqslant i$ and $i_2 \leqslant i$. (In practice, $(I, \leqslant)$ will often be a linearly ordered set.)

An *inverse system* of groups over a directed set I is a family of groups $\{G_i : i \in I\}$ together with connecting homomorphisms $\pi_{ij} : G_i \longrightarrow G_j$ whenever $i \geqslant j$ satisfying the compatibility conditions: π_{ii} is the identity on G_i and $\pi_{ij}\pi_{jk} = \pi_{ik}$ whenever $i \geqslant j \geqslant k$.

The *inverse limit* $\varprojlim G_i$ of such an inverse system $\{G_i : i \in I\}$ is the subgroup of the cartesian product $\prod_{i \in I} G_i$ consisting of all (g_i) with the property that $g_i \pi_{ij} = g_j$ whenever $i \geqslant j$.

1.2 Pro-p groups

A *topological group* is a group with the underlying structure of a topological space with respect to which the group operations are continuous. We shall view finite groups as topological groups with the discrete topology. Let $\{G_i : i \in I\}$ be an inverse system of *finite* groups. The inverse limit of such a system is called a *profinite group* and has the structure of a topological group where the topology on $\varprojlim G_i$ is induced from the product topology on $\prod_{i \in I} G_i$. The following theorem gives a topological characterization of profinite groups.

Theorem 1.1. *A topological group is profinite if, and only if, it is a compact Hausdorff topological group having a neighbourhood base for the identity consisting of open subgroups.* (See [7] Chapter 1.)

We now intoduce our leading character. Let p be a prime. A *pro-p group* is an inverse limit of finite p-groups. We have an analogous result to Theorem 1.1.

Theorem 1.2. *A topological group is a pro-p group if, and only if, it is a profinite group in which every open normal subgroup has index equal to a power of p.* (See [7] Chapter 1.)

Example 1.3. Profinite and pro-p groups

(i) Every finite group is an instance of a profinite group. A discrete group is a pro-p group if, and only if, it is a finite p-group.

(ii) The simplest non-discrete example of a pro-p group is the additive group of p-adic integers

$$\mathbb{Z}_p = \varprojlim \mathbb{Z}/p^i\mathbb{Z}.$$

This group is the historical starting point for the theory of pro-p groups and became the prototype for all pro-p groups. It plays the rôle in the theory of pro-p groups analogous to the infinite cyclic group in abstract group theory. We shall see later that the 'analytic' pro-p groups are built from finitely many copies of $\mathbb{Z}_p$. This depends on the fact that any pro-p group G admits a natural action of $\mathbb{Z}_p$: if $\lambda = \lim_{n\longrightarrow\infty} a_n$ is a p-adic integer where $a_n \in \mathbb{Z}$ for each n, we define

$$x^\lambda = \lim_{n\longrightarrow\infty} x^{a_n}.$$

It is an easy exercise to show that this is well-defined.

(iii) We can generalize the construction of the p-adic integers to associate a pro-p group with any abstract group. Let Γ be a group and $\{N_i : i \in I\}$ a system of normal subgroups of p-power index in Γ with the property that, given any N_i and N_j, there exists $N_k \subseteq N_i \cap N_j$. If we partially order I by defining $i \leqslant j$ whenever $N_j \subseteq N_i$, then $(I, \leqslant)$ is a directed set. Then $\{\Gamma/N_i : i \in I\}$ becomes an inverse system of finite p-groups with connecting homomorphisms given by the natural maps $\Gamma/N_j \longrightarrow \Gamma/N_i$. The inverse limit $G = \varprojlim \Gamma/N_i$ is then a pro-p group. If $\{N_i : i \in I\}$ is the set of all normal subgroups of p-power index in Γ, we call G the *pro-p completion of* Γ, denoted by $\hat{\Gamma}_p$. For example, $\mathbb{Z}_p$ (which we write instead of $\hat{\mathbb{Z}}_p$) is the pro-p completion of $\mathbb{Z}$. We should also mention the *profinite completion of* Γ, denoted by $\hat{\Gamma}$, which is constructed by taking the inverse limit of all finite quotients of Γ. $\hat{\Gamma}_p$ is then the maximal pro-p quotient of $\hat{\Gamma}$. We leave it as an exercise to prove that $\hat{\mathbb{Z}}$, the profinite completion of $\mathbb{Z}$, is isomorphic to $\prod_p \mathbb{Z}_p$.

(iv) The group $G = \mathrm{GL}_d(\mathbb{Z}_p)$ of all d by d invertible matrices over $\mathbb{Z}_p$ is a profinite group with a base of neighbourhoods of 1 in G given by the *congruence subgroups* $G_i = \{g \in G : g \equiv 1_d \bmod p^i\}$. The open subgroup G_1 is a pro-p group.

It is Example 1.3(iii) which provides us with the passage from abstract groups to pro-p groups. We shall be especially interested in groups Γ for which there is a system $\{N_i : i \in I\}$ of normal subgroups of p-power index in Γ with the property that $\bigcap_{i\in I} N_i = 1$. Such groups are called *residually finite-p*. In this situation we can see Γ sitting naturally as a dense subset in $G = \varprojlim \Gamma/N_i$ via the diagonal injection $\Gamma \longrightarrow G$ defined by $g \longmapsto (gN_i)$. In Section 2 we shall see that, by realizing Γ as a subgroup of G, we can exploit the extra structure available to us on the topological group G to prove results about Γ. (We refer the reader to [8] and [9] for further examples of this philosophy.) We describe first a class of pro-p groups whose structure is particularly rich.

1.3 Analytic pro-p groups

We call a pro-p group an *analytic pro-p group* if it is a Lie group whose local coordinate systems are p-adic valued rather than real valued. Given such a group we have all the usual apparatus of Lie theory available to us from which we can deduce some special properties of this class of pro-p groups.

The reader may feel a little unsatisfied (or relieved) by our curt definition of analytic pro-p groups. However, Lazard's great achievement was to show that this class of groups has a straightforward group-theoretic characterization, thereby solving the p-adic version of Hilbert's fifth problem. This group-theoretic condition on our pro-p group is in fact sufficient to construct internally the special architectural features of this class of pro-p groups without ever introducing the imposing edifice of Lie theory. We thus proceed to outline Lazard's criterion together with a more recent characterization due to Lubotzky and Mann.

Theorem 1.4. *Let G be a pro-p group. Then G is an analytic pro-p group if, and only if, (i) G is topologically finitely generated (i.e. there exists a finite set of elements which generate a dense subgroup of G); and (ii) G contains an open subgroup H with the property that H/H^p (or H/H^4, if $p = 2$) is abelian (where G^n denotes the closure of the subgroup generated by the nth powers of G).* (See [21] and [7].)

Lubotzky and Mann call such a group H a *powerful pro-p group*. It is striking that knowledge of a small finite quotient of G should so globally affect the structure of the group. We give a hint in Section 1.4 as to why this should give rise to both an underlying manifold structure on G and a natural Lie algebra associated with G. Lubotzky and Mann study powerful pro-p groups in two papers [25] and [26]. Their work culminates in a second characterization in terms of rank which turns out to be more relevant to applications.

The *rank* of an abstract group Γ is defined to be

$$r(\Gamma) = \sup\{d(H): H \leqslant \Gamma\}$$

where $d(H)$ = minimal number of generators of H. We define the *rank* of a pro-p group G to be equal to

$$r(G) = \sup\{r(G/N) : N \text{ an open normal subgroup of } G\}.$$

(This definition of the rank of a pro-p group more readily applies to our situation than the more natural, but equivalent definition

$$r(G) = \sup\{d(H) : H \text{ an open subgroup of } G\},$$

where $d(H)$ is the minimal number of topological generators of H.)

Theorem 1.5. *Let G be a pro-p group. Then G is an analytic pro-p group if, and only if, G has finite rank.* (See [26] and [7].)

Example 1.6. Analytic pro-p groups

(i) Every finite p-group is an analytic pro-p group.

(ii) The additive group of p-adic integers is an analytic pro-p group. So too is any topologically finitely generated abelian pro-p group.

(iii) As pointed out above, it is Example 1.3(iii) which is going to be the class of pro-p groups of most interest to us. If we want to exploit the special properties of the class of analytic pro-p groups we will need to establish when the pro-p completion of an abstract group Γ is analytic. Let us mention one condition which guarantees that this is so for each prime p. Define the upper rank of a group Γ to be

$$ur(\Gamma) = \sup\{r(\Gamma/N) : N \text{ a normal subgroup of finite index in } \Gamma\}.$$

Theorem 1.5 implies that if Γ has finite upper rank, then $\hat{\Gamma}_p$ is analytic for all primes p. (Note that, unlike the case of pro-p groups, upper rank and rank are in general distinct for abstract groups.)

(iv) We call a profinite group a *compact p-adic analytic group* if it contains an open analytic pro-p group. The profinite group $G = \mathrm{GL}_d(\mathbb{Z}_p)$ is an example of a compact p-adic analytic group since G_1, the first congruence subgroup of G, is an analytic pro-p group. The fact that G_1 is analytic follows from Theorem 1.5 together with the fact that every finite quotient of G has rank bounded by d^2. Alternatively, G_1 has a natural global p-adic coordinate system with respect to which group multiplication is given by polynomial functions. This underlying p-adic manifold structure is compatible with the topology defined by the congruence subgroups G_i and makes G_1 into an analytic group.

We will give later a third characterization of the class of analytic pro-p groups due to Lubotzky and Mann in terms of arithmetic data associated with the lattice of subgroups of a pro-p group. Further characterizations may be found in [7].

1.4 Some properties of analytic pro-p groups

We indicate in this section how we can deduce some of the more Lie-theoretic properties of the class of analytic pro-p groups from the group-theoretic characterization of Lazard. A more detailed account of this algebraic approach to analytic pro-p groups may be found in [7].

We take as our starting point a (topologically) finitely generated pro-p group G. We define the *lower p-series* associated with G by $G_1 = G$ and, for $i \geqslant 1$, $G_{i+1} = G_i^p[G_i, G]$ (where $[G_i, G]$ denotes the closure of the group generated by commutators in G_i and G). Note that, as for finite p-groups, $d(G) = d(G/G_2)$. If we assume in addition that G is powerful, then the mapping $x \longmapsto x^p$ induces epimorphisms $P_i: G_i/G_{i+1} \longrightarrow G_{i+1}/G_{i+2}$. Since G_i/G_{i+1} is finite, there exists m such that P_i becomes an isomorphism for all $i \geqslant m$. The group G_m is called a ***uniformly powerful pro-p group***, that is a finitely generated powerful pro-p group whose factors in the lower central p-series are all isomorphic. (Lazard refers to these groups as *p-saturable*.) In Example 1.3(iv), $G = \mathrm{GL}_d(\mathbb{Z}_p)$, it can be shown that the principal congruence subgroup modulo p if p is odd (respectively modulo 4 if $p = 2$) is a uniformly powerful pro-p group. This is a consequence of a simple instance of Hensel's Lemma (see [7] Chapter 5).

We focus our attention now on a uniformly powerful pro-p group G. It is possible to identify a very natural underlying manifold structure on G. Let $r = d(G)$; then $G_i/G_{i+1} \cong \mathbb{F}_p^r$ for all $i \geqslant 1$. Let $x_1, \ldots, x_r$ generate G_1 modulo G_2. We then have the following theorem (recall the definition of the action of $\mathbb{Z}_p$ on G given in Example 1.3(ii)).

Lemma 1.7. *Let $x \in G$. There exist unique p-adic integers $\lambda_1, \ldots, \lambda_r \in \mathbb{Z}_p$ such that*

$$x = x_1^{\lambda_1} \ldots x_r^{\lambda_r}.$$

We may thus define a homeomorphism $\varphi: G \longrightarrow \mathbb{Z}_p^r$ via $x \longmapsto (\lambda_1, \ldots, \lambda_r)$.

Proof It suffices to define, for each $i = 1, \ldots, r$ and $j \geqslant 1$, $\lambda_i(j) \in \mathbb{Z}$ such that $\lambda_i(j) - \lambda_i(j+1) \in p^j\mathbb{Z}$ and

$$x \equiv x_1^{\lambda_1(j)} \ldots x_r^{\lambda_r(j)} \bmod G_{j+1}.$$

We define $\lambda_i(j)$ recursively. Note that for $k \geqslant 1$, if we set $q = p^{k-1}$, then $x_1^q, \ldots, x_r^q$ generate G_k modulo G_{k+1}. This provides our base step and the key to our recursion. Suppose that we have defined $\lambda_i(j)$ for $1 \leqslant j < k$; then

$$x \equiv x_1^{\lambda_1(k-1)} \dots x_r^{\lambda_r(k-1)} \bmod G_k.$$

So, for some $g_k \in G_k$,

$$x \equiv x_1^{\lambda_1(k-1)} \dots x_r^{\lambda_r(k-1)}.g_k \bmod G_{k+1}.$$

Now there exist $\mu_1, \dots, \mu_r \in \{0, \dots, p-1\}$ such that $g_k \equiv x_1^{q\mu_1} \dots x_r^{q\mu_r} \bmod G_{k+1}$. But note that G_k/G_{k+1} is central in G/G_{k+1} by definition of the lower p-series. So, setting $\lambda_i(k) = \lambda_i(k-1) + q\mu_i$, we deduce that

$$x \equiv x_1^{\lambda_1(k)} \dots x_r^{\lambda_r(k)} \bmod G_{k+1}.$$

Thus setting $\lambda_i = \lim_{j \longrightarrow \infty} \lambda_i(j)$, we have $x = x_1^{\lambda_1} \dots x_r^{\lambda_r}$. We leave the uniqueness of this expression as an exercise. ◊ ◊

Lazard then went on to show that the group operations with respect to this coordinate system are given by analytic functions. To do this he considers a completion $\hat{R}$ of the group ring $R = \mathbb{Z}_p[G]$ with respect to a filtration induced from the lower p-series on G. It is also in the context of this completion $\hat{R}$ that we can identify a natural Lie algebra L associated with G. There is a close link between representations of the Lie algebra L and of the group G which enables us to translate Ado's theorem for finite dimensional Lie algebras (i.e. the existence of a faithful finite-dimensional representation of L) into the following:

Theorem 1.8. *If G is a uniformly powerful pro-p group, then there exists a faithful representation $\varphi: G \longrightarrow \mathrm{GL}_d(\mathbb{Z}_p)$.*

Every analytic pro-p group has a subgroup of finite index which is a uniformly powerful pro-p group. Thus, inducing the representation provided by Theorem 1.8, we deduce the following property of analytic pro-p groups, which we shall use many times in Section 2:

Theorem 1.9. *If G is an analytic pro-p group, then there exists a faithful representation $\varphi: G \longrightarrow \mathrm{GL}_d(\mathbb{Z}_p)$.*

Although we cannot hope to present here the details behind these remarks (see [7] Chapter 8), it is a remarkable fact that we can prove, without any extra machinery, a weaker form of Theorem 1.9 which is sufficient for most applications. This depends on the possibility of defining a Lie algebra structure directly onto a uniformly powerful pro-p group G. The mapping $x \longmapsto x^{p^n}$ gives rise to a bijection between G and G_{n+1}; we will denote its inverse by $x \longmapsto x^{p^{-n}}$. We now define two binary operations $+$ and $(\ ,\ )$ on G as follows:

$$\begin{aligned} x + y &= \lim_{n \longrightarrow \infty} (x^{p^n} y^{p^n})^{p^{-n}} \\ (x, y) &= \lim_{n \longrightarrow \infty} [x^{p^n}, y^{p^n}]^{p^{-2n}}. \end{aligned}$$

Setting $\lambda . x = x^\lambda$ for $\lambda \in \mathbb{Z}_p$, we then have:

Theorem 1.10. *Let G be an r-generator uniformly powerful pro-p group. Then $(G, +, (\ ,\ ))$ is a free $\mathbb{Z}_p$-Lie algebra of rank r.* (See [7].)

The action of $\mathrm{Aut}(G)$ on this $\mathbb{Z}_p$-Lie algebra $(G, +, (\ ,\ ))$ induces a faithful representation $\mathrm{Aut}(G) \longrightarrow \mathrm{GL}_r(\mathbb{Z}_p)$. Hence, via the inner automorphisms, we have a representation $G \longrightarrow \mathrm{GL}_r(\mathbb{Z}_p)$ with kernel equal to the centre of G. By inducing this representation we thus have the following weaker version of Theorem 1.9:

Theorem 1.11. *If G is an analytic pro-p group, then there is a representation $\varphi: G \longrightarrow \mathrm{GL}_d(\mathbb{Z}_p)$ with abelian kernel.* (See [7].)

Lazard was also interested in cohomological properties of the class of analytic pro-p groups. He used the embedding of G and its Lie algebra L into the completion $\hat{R}$ of $R = \mathbb{Z}_p[G]$ to deduce that the cohomology of G as a pro-p group can be identified with the cohomology of the Lie algebra L. The cohomological dimension of an r-generator uniformly powerful pro-p group is r. In fact such groups are Poincaré duality groups of dimension r. Those interested in cohomological questions should consult Chapter 5 of Lazard's paper [21].

We shall also be interested in the rather well-behaved combinatorial aspects of analytic pro-p groups. The combinatorial group theory of pro-p groups in general turns out to be somewhat easier than for abstract groups. (See [22] and [47].)

If $\Gamma = F_e$, the free group on e generators, let $\hat{F}_e(p) = \hat{\Gamma}_p$, the pro-$p$ completion of Γ. Then, if $d \leqslant e$, every pro-p group G generated topologically by d elements arises as a quotient of $\hat{F}_e(p)$. We call $\hat{F}_e(p)$ the free pro-p group on e generators. A *presentation* of G is then a pair $\langle X; R \rangle$ where X is a finite set with $|X| = e$, and R is a subset of $\hat{F}_e(p)$ with the property that G is isomorphic to $\hat{F}_e(p)/N$ where N is the minimal closed normal subgroup of $\hat{F}_e(p)$ containing R. A presentation $\langle X; R \rangle$ for G is called *minimal* if $|X| = d(G)$. For the class of analytic pro-p groups we have the following relationship between $|X|$ and $|R|$ for a minimal presentation (see [23] and [7] Chapter 6):

Theorem 1.12. *Let G be an analytic pro-p group with a minimal presentation $\langle X; R\rangle$. If G is not isomorphic to $\mathbb{Z}_p$, then*

$$|R| \geqslant \frac{|X|^2}{4}.$$

This is an extension of the theorem of Golod and Šafarevič for finite p-groups. We shall see a rather striking application of this result to the Congruence Subgroup Problem in Section 2.

2 Applications of analytic pro-p groups

We present in this section a selection of results which have been proved using the theory of analytic pro-p groups. Underlying the diverse range of results is a unifying methodology. We reduce to the case of a residually finite-p group Γ, which can be embedded into a pro-p group constructed from some inverse limit of finite p-quotients of Γ. We then hope to identify some property of Γ which implies that this pro-p group is analytic, allowing us to plunder the results of Section 1.4.

The first application we give is to combinatorial group theory for abstract groups.

2.1 Presentations of abstract groups

As we pointed out in Section 1.4, combinatorial group theory for abstract groups is in general harder than for pro-p groups. However, the following fact often allows us to reduce the former to the latter. If $\langle X; R\rangle$ is a presentation for the finitely generated group Γ, then $\langle X; R\rangle$, considered as a presentation for a pro-p group, is actually a presentation for the pro-p completion $\hat{\Gamma}_p$ of Γ. Thus, if this pro-p completion is analytic we might hope to translate our Golod–Šafarevič inequality to the presentation of Γ. However, a minimal presentation for Γ does not in general give rise to a minimal presentation for the pro-p completion. We still have a variant which involves the following invariant

$$d_p(\Gamma) = \dim_{\mathbb{F}_p}(\Gamma/[\Gamma,\Gamma]\Gamma^p) = d(\hat{\Gamma}_p)$$

(see [23]).

Theorem 2.1. *Let Γ be a finitely generated group with a presentation $\langle X; R\rangle$. If $\hat{\Gamma}_p$ is analytic and not isomorphic to $\mathbb{Z}_p$, then*

$$|R| - (|X| - d_p(\Gamma)) \geqslant \frac{d_p(\Gamma)^2}{4}.$$

For example, we saw in Section 1.3 that if Γ is a group of finite upper rank then the pro-p completion of Γ is analytic for all primes p. Such groups include *polycyclic groups*, that is, groups Γ with a finite series of subgroups $1 \lhd N_1 \lhd \cdots \lhd N_k = \Gamma$ with the property that N_{i+1}/N_i is cyclic for all i. However, for the subfamily of finitely generated nilpotent groups we can find a prime p with the property that $d_p(\Gamma)$ = minimal number of generators of Γ. We thus have a direct analogue of the Golod–Šafarevič inequality (see [23]):

Corollary 2.2. *Let Γ be a finitely generated, nilpotent group with a minimal presentation $\langle X; R\rangle$. Then, if Γ is not isomorphic to $\mathbb{Z}$,*

$$|R| \geqslant \frac{|X|^2}{4}.$$

It is interesting to note that the original context of the Golod–Šafarevič inequality for finite groups was in class field theory. It was used by Šafarevič (together with Golod) to show that in general the Hilbert class field tower can be infinite (see [43]). More recently, in a different vein, Sen has considered totally ramified Galois extensions of a local field K (of characteristic 0 and residue characteristic p) whose Galois groups G are analytic pro-p groups, establishing a conjecture due to Serre connecting the lower p-series of G with the filtration given by ramification subgroups in the 'upper numbering' (see [46]).

In the next section we shall describe a remarkable application of Theorem 2.1 made by Lubotzky to the Congruence Subgroup Problem.

2.2 Congruence Subgroup Problem

Let $\Gamma = \mathrm{SL}_n(\mathbb{Z})$. For each $q \in \mathbb{N}$, we set $\Gamma_q = \{x \in \Gamma : x \equiv 1_n \bmod q\mathbb{Z}\}$ which is the kernel of the map $\pi_q\colon \mathrm{SL}_n(\mathbb{Z}) \longrightarrow \mathrm{SL}_n(\mathbb{Z}/q\mathbb{Z})$. We call a subgroup of Γ a *congruence subgroup* if it contains Γ_q for some $q \in \mathbb{N}$. Congruence subgroups clearly have finite index in Γ since $\mathrm{SL}_n(\mathbb{Z}/q\mathbb{Z})$ is a finite group. Conversely, we may pose the following question.

Question 2.3. (The Congruence Subgroup Problem) *Is every subgroup of finite index in Γ a congruence subgroup?*

The answer is in fact 'yes' when $n \geqslant 3$ (see [3] and [2]); but when $n = 2$, $\mathrm{SL}_2(\mathbb{Z})$ has a vast assortment of normal subgroups of finite index only a few of which are congruence subgroups (see [13] and, more recently, [19]).

We can generalize the Congruence Subgroup Problem. Let k be a finite extension of $\mathbb{Q}$ and let G be a linear algebraic group defined over k. Let S be a finite set of valuations on k containing all the archimedean valuations, and denote by ϑ^S the ring of S-integers, that is

$$\vartheta^S = \{x \in k : \nu(x) \geqslant 0 \text{ for all } \nu \notin S\}.$$

(For example, if $k = \mathbb{Q}$ and $S = \{\nu_\infty, \nu_p\}$ where ν_∞ is the archimedean valuation on $\mathbb{Q}$ and ν_p is the p-adic valuation, then $\vartheta^S = \mathbb{Z}[1/p]$.) We write $\vartheta^S = \vartheta$ if S is just the set of archimedean valuations.

We fix a faithful k-rational representation $\rho\colon G(k) \longrightarrow \mathrm{GL}_n(k)$. Let $\Gamma = \rho^{-1}(\rho G(k) \cap \mathrm{GL}_n(\vartheta^S))$. A subgroup of finite index in Γ is called an *S-congruence subgroup* if it contains a subgroup $\Gamma_a = \{x \in \Gamma : \rho(x) \equiv 1_n \bmod a\}$ where a is a non-zero ideal in ϑ^S. The Congruence Subgroup

Problem then becomes the question of whether every subgroup of finite index in Γ is an S-congruence subgroup.

We can reformulate this problem in terms of profinite groups. Let $\bar{\Gamma}$ be the inverse limit of the system of finite subgroups

$$\{\Gamma/\Gamma_a : a \text{ is a non-zero ideal in } \vartheta^S\}$$

and $\hat{\Gamma}$ the profinite completion of Γ. We have a natural surjective map $\pi : \hat{\Gamma} \longrightarrow \bar{\Gamma}$. The Congruence Subgroup Problem is then equivalent to the question of whether the kernel of π is trivial.

It is not difficult to show that this problem is dependent only on the algebraic group G and the set of valuations S of k. In fact Serre conjectured in [48] that the finiteness of $\ker \pi$ is dependent only on the Lie group $G(k_\nu)$ where k_ν is the completion of k with respect to $\nu \in S$. More precisely, if k_ν-rank(G) denotes the dimension of a maximal torus of the k_ν-Lie group $G(k_\nu)$, then we have:

Conjecture 2.4. (The Congruence Subgroup Conjecture) $\ker \pi$ *is finite if, and only if,* $\sum_{\nu \in S} k_\nu$-rank$(G) \geqslant 2$.

G is said to have the congruence subgroup property (CSP) if $\ker \pi$ is finite. That the CSP for G should follow from the condition on the k_ν-ranks of G has to a large extent been proved and we refer the reader to [39] and [40] for the precise results. Less is known about the case now where Card$(S) = 1$ and k_ν-rank$(G) = 1$. However, we shall describe below a result which presented a new approach to contradicting the CSP.

We write

$$G(\widehat{\vartheta^S}) = \prod_{\nu \notin S} G(\vartheta_\nu)$$

where ϑ_ν is the ring of integers of k_ν. For example, if $k = \mathbb{Q}$ and $S = \{\nu_\infty\}$ then

$$G(\hat{\mathbb{Z}}) = \prod_p G(\mathbb{Z}_p).$$

Γ embeds naturally into $G(\widehat{\vartheta^S})$ and it is an easy exercise to prove that its closure is isomorphic to $\bar{\Gamma}$. The algebraic group G is said to have *strong approximation* if in fact Γ is dense in $G(\widehat{\vartheta^S})$. Kneser [20] conjectured that various necessary conditions for G to have strong approximation were in fact sufficient. His conjecture was later established by Platonov (see [36] and [37]). Our interest in strong approximation is the following result:

Theorem 2.5. *Assume that G has strong approximation and that H is a subgroup of finite index in Γ. If G has the CSP, then, for each prime p, the pro-p completion $\hat{H}_p$ of H is analytic.*

Proof We shall outline a proof in the situation $G = \mathrm{SL}_n$ where $n \geqslant 3$, $S = \{\nu_\infty\}$ and $\Gamma = \mathrm{SL}_n(\mathbb{Z})$. SL_n has strong approximation, a fact that can be proved by choosing a suitable set of generators for $\mathrm{SL}_n(\mathbb{Z})$. Since $\ker\pi$ is in fact trivial in our situation we have that $\hat{\Gamma} \cong \prod_q \mathrm{SL}_n(\mathbb{Z}_q)$. We can therefore identify $\hat{H}$ with an open subgroup of $\prod_q \mathrm{SL}_n(\mathbb{Z}_q)$. Without loss of generality $\hat{H} = \prod_q U_q$ where U_q is an open subgroup of $\mathrm{SL}_n(\mathbb{Z}_q)$. $\hat{H}_p$ is the maximal pro-p quotient of $\hat{H}$. So $\hat{H}_p = \prod_q L_q$ where L_q is a pro-p quotient of U_q. Every finite quotient of $\mathrm{SL}_n(\mathbb{Z}_q)$ has bounded rank; therefore the pro-p group L_q has finite rank.

By a theorem of Borel and Harish–Chandra [5],we have that Γ, and hence H, is finitely generated. Since $d(\hat{H}_p) = \sum_q d(L_q)$ it follows that $L_q = 1$ for almost all q. So $\hat{H}_p = \prod_q L_q$ has finite rank and, by Theorem 2.5, is p-adic analytic. ◊ ◊

Contrast this with $\mathrm{SL}_2(\mathbb{Z})$ which contains a subgroup of finite index whose pro-p completion is a non-abelian free pro-p group.

Lubotzky was able to employ Theorem 2.5 in the following situation:

Theorem 2.6. *Let G be a semisimple algebraic group over k with strong approximation, and Γ a subgroup of finite index in $G(\vartheta)$ with the property that Γ can be embedded in* $\mathrm{SL}_2(\mathbb{C})$ *as a discrete subgroup. Then G does not have the CSP.*

The structure of discrete subgroups in Lie groups is rather mysterious (see [18] and [38]). Previous techniques to prove such a theorem had reduced the problem to a conjecture of Thurston about hyperbolic 3-manifolds which is still open. However, Lubotzky's approach exploits a theorem due to Epstein (see [12] and [41]), again a result in the context of 3-manifolds, which says that for a torsion-free, finitely generated discrete subgroup Δ of $\mathrm{SL}_2(\mathbb{C})$

$$\sup\{|X| - |R| : \langle X; R\rangle \text{ is a presentation for } \Delta\} \geqslant 0.$$

If the algebraic group of Theorem 2.6 had the CSP, then, by Theorem 2.5, subgroups of finite index in Γ would satisfy the variant of the Golod–Šafarevič inequality detailed in Theorem 2.1. Combining this with the inequality of Epstein's theorem, it is possible to bound $d_p(\Delta)$ for all subgroups of finite index in Γ and all primes p. Taking $p = 2$ and applying the results of Mal'cev on linear groups we can conclude that Γ is soluble-by-finite. By the Density Theorem of Borel [4] this forces G to be soluble-by-finite, contradicting the hypothesis that G is semisimple. We refer the reader to Section 6.6 in [7] for a more detailed exposition of the proofs of Theorem 2.5 and Theorem 2.6.

Bass et al. conversely use Theorem 2.5 to prove rigidity results for various arithmetic groups (see [3]). For example, they use their solution to the CSP for $\Gamma = \mathrm{SL}_n(\mathbb{Z})$ $(n \geqslant 3)$ together with Theorem 2.5 to deduce that all representations of Γ are in fact polynomial. The key to this result is essentially the relationship between representations of an analytic group and its Lie algebra. However, also lurking behind this result are various arguments in the theory of algebraic groups.

2.3 Characterization of linear groups

We turn now to a famous old problem in group theory, namely to give a group-theoretic criterion to determine when a group Γ is linear. We restrict ourselves to the characterization of finitely generated linear groups over a field of characteristic 0. There are clearly many necessary conditions: for example, Γ should be residually finite, virtually torsion-free, and, by the celebrated theorem of Tits [49], if not soluble-by-finite, then it should have a free non-abelian subgroup. However, none of these properties suffice to characterize such groups. The problem is similar in nature to the problem of characterizing pro-p groups with an analytic structure. Lubotzky showed that for linear groups over a field of characteristic 0, the one can be reduced to the other.

Theorem 2.7. *Let Γ be a group. Suppose that Γ has a descending chain of normal subgroups $\{N_i : i \in \mathbb{N}\}$ such that*

(i) $|\Gamma : N_1|$ is finite;

(ii) for some prime p, N_1/N_i is a finite p-group for every $i \geqslant 1$;

(iii) $\bigcap_{i \in \mathbb{N}} N_i = \{1\}$;

(iv) there exists $c \in \mathbb{N}$ such that $d(N_i/N_j) \leqslant c$ for all $j \geqslant i \geqslant 1$.

Then Γ has a faithful linear representation over a field of characteristic 0. Conversely, if Γ is a finitely generated linear group over a field of characteristic 0, then Γ has such a chain $\{N_i : i \in \mathbb{N}\}$. (See [24].)

The converse is established by showing that the linear group Γ has a faithful linear representation over $\mathbb{Z}_p$, for some prime p, and taking the congruence subgroups modulo powers of p.

However, we are interested in why the existence of such a chain of subgroups should imply that Γ is linear. Note that it suffices to prove the linearity of N_1 since we can then induce a faithful linear representation from N_1 to Γ. Conditions (ii) and (iii) imply that N_1 embeds into the pro-p group

$$G = \varprojlim N_1/N_i.$$

Finally, it is shown in [24] that condition (iv) implies that G has finite rank. By Theorem 1.5 and Theorem 1.9, we may therefore infer the linearity of G and thus of N_1.

In practice it is as difficult to establish the existence of such a system of normal subgroups for a specific group (such as the Artin braid group, the linearity of which is still an open problem) as to show that the group is linear. The characterization of finitely generated linear groups over fields of finite characteristic is still open. So too is the problem of characterizing linear groups which are not necessarily finitely generated. This might seem surprising to those with a knowledge of Mal'cev's result that a group Γ has a faithful linear representation of degree d over a field of characteristic 0 if, and only if, every finitely generated subgroup of Γ has such a representation of degree d. However, Lubotzky's result fails to characterize finitely generated groups of a given degree.

Theorem 2.7 does make an important appearance in the theorem of the next section on groups satisfying various finiteness conditions.

2.4 Residually finite groups

We call a group G *residually* $\wp$ for some property $\wp$ of groups (e.g., finite, nilpotent, etc.) if

$$\bigcap\{N : N \lhd G \text{ and } G/N \text{ has } \wp\} = 1;$$

we call G *virtually-*$\wp$ if G has a subgroup H of finite index and H has $\wp$.

Theorem 2.8. *A finitely generated, residually finite group of finite rank is virtually-soluble.* (See [27].)

This beautiful theorem depends on a series of reductions, the first of which involves some deep theorems in finite groups, for example the *Odd Order Theorem* that every finite group of odd order is soluble. Such a theorem allows us to reduce to the case of a residually finite-soluble group. The second reduction applies results on finitely generated soluble groups of finite rank. Attention then focuses on residually finite-p groups. We are now in a position to apply Lubotzky's theorem 2.7 to reduce to linear groups. An earlier result of Platonov on linear groups of finite rank then suffices to complete the proof. Note that finitely generated soluble groups of finite rank are the soluble *minimax* groups, as proved by Robinson [42].

We refer the reader to [32] for various generalizations of this theorem.

2.5 Growth of finite index subgroups

Let Γ be a group and denote by $a_n(\Gamma)$ the number of subgroups of index n in Γ. This is an invariant associated with the class of residually finite groups since it only sees as far as the maximal residually finite quotient of any group. We are interested in groups for which this function is finite valued, which is certainly so if we restrict ourselves to the class of finitely generated groups. (In fact, this is true if we consider groups of finite upper rank [32].) In [44], Segal posed the problem of characterizing those finitely generated, residually finite groups Γ for which the function $a_n(\Gamma)$ grows polynomially (i.e. there exist $c, s \in \mathbb{N}$ such that $a_n(\Gamma) \leqslant cn^s$ for all $n \in \mathbb{N}$). Such groups we shall call groups with PSG. In the same paper Segal gave a partial answer, proving that a finitely generated residually nilpotent soluble group has PSG if, and only if, it is soluble of finite rank. However, Lubotzky and Mann prove the same theorem without assuming solubility by making use of a third characterization of analytic pro-p groups.

Theorem 2.9. *A finitely generated, residually nilpotent group has PSG if, and only if, it is a soluble group of finite rank.* (See [28].)

We first reduce to the case of a residually finite-p group Γ which can then be embedded in its pro-p completion $G = \hat{\Gamma}_p$. Let $c_n(G)$ denote the number of open subgroups of G of index n. Then $c_n(G) \leqslant a_n(\Gamma)$ for all $n \in \mathbb{N}$. We now exploit the following characterization of analytic pro-p groups to reduce to the case of finitely generated linear groups.

Theorem 2.10. *Let G be a (topologically) finitely generated pro-p group. Then G is analytic if, and only if, $c_n(G)$ grows polynomially.* (See [28].)

Theorem 2.9 is now a consequence of the following theorem together with Segal's theorem.

Theorem 2.11. *A finitely generated linear group Γ with PSG is virtually-soluble.* (See [28].)

Let us recall a different growth function studied by Milnor, Wolf and Bass associated with the class of finitely generated groups (see [34], [51] and [1]). Let $x_1, \ldots, x_r$ be a set of generators for Γ and let $f_n(\Gamma)$ be the number of elements of Γ which can be expressed by a word of length $k \leqslant n$ in this set of generators. It is a deep theorem due to Gromov [15] that $f_n(\Gamma)$ grows polynomially if, and only if, Γ is virtually-nilpotent. It is interesting to observe that Gromov also reduces to the case of linear groups but makes use instead of the characterization of real Lie groups. We mention as a curiosity that we can effect this reduction much more swiftly for the subclass of finitely generated, residually nilpotent groups by showing that the pro-p completion of Γ has finite rank if $f_n(\Gamma)$ grows polynomially. In fact Grigorchuk has shown, using properties of the graded algebra of an analytic pro-p group, that this is also true under the weaker hypothesis that $f_n(\Gamma) < \exp(\sqrt{n})$ for all $n \geqslant 1$ (see [14]).

Although the proof of Theorem 2.11 is independent of the methods expounded in this chapter, it seems a shame not to give at least a taste of the ingredients. Assuming Γ is not virtually-soluble, we can apply the results of Jordan, Zassenhaus, Mal'cev and Platonov to reduce to a finitely generated linear group over the ring $R = \mathbb{Z}[1/m]$ for some $m \in \mathbb{N}$. A recent variant of the Strong Approximation Theorem proved by Nori [35] and Weisfeiler [33] ensures that if the Zariski closure $\bar{\Gamma}$ of $\Gamma \leqslant \mathrm{GL}_d(R)$ is a simply connected, semisimple algebraic group then the closure of Γ with respect to the congruence topology on $\mathrm{GL}_d(R)$ has finite index in $\bar{\Gamma}(R) = \bar{\Gamma} \cap \mathrm{GL}_d(R)$. So if Γ has PSG then $\bar{\Gamma}(R)$ has PSG. However, a cunning application of a weak form of the *Prime Number Theorem* shows that the rate of growth of the number of congruence subgroups in $\bar{\Gamma}(R)$ is faster than polynomial.

Let us conclude this section by mentioning the following refinement of Theorem 2.9 which represents the best answer to date to Segal's problem.

Theorem 2.12. *A finitely generated, residually finite-soluble group has PSG if, and only if, it is a soluble group of finite rank.* (See [32].)

At present the reduction from residually finite groups to residually finite-soluble groups is proving rather elusive and, just as in Theorem 2.8, seems likely to involve deep theorems in finite group theory (e.g., *the Classification of Finite Simple Groups*).
[**Note added in proof:** this final reduction has now been achieved (see [7] Chapter 6).]

2.6 Poincaré series

Most of the previous results have exploited the passage through analytic pro-p groups to reduce to the class of linear groups. We describe in this section a result which depends more on the uniform architectural features of analytic pro-p groups. The result relates to the following Poincaré series we can associate with a finitely generated group Γ, for each prime p:

$$\zeta_{\Gamma,p}(s) = \sum_{n=0}^{\infty} a_{p^n}(\Gamma)p^{-ns}.$$

These are the local factors associated with the global zeta function:

$$\zeta_{\Gamma}(s) = \sum_{n=1}^{\infty} a_n(\Gamma)n^{-s}.$$

There are various natural questions we can ask about these number theoretical functions associated with Γ: the existence of an Euler product for $\zeta_\Gamma(s)$ in terms of its local factors $\zeta_{\Gamma,p}(s)$; the relationship between the poles of $\zeta_\Gamma(s)$ and the group Γ; the existence of any functional equations for $\zeta_\Gamma(s)$. However, we restrict ourselves to the following:

Question 2.13. *For which Γ is $\zeta_{\Gamma,p}(s)$ a rational function in p^{-s}?*

Example 2.14.

(i) If $\Gamma = \mathbb{Z}^d$, then

$$\zeta_{\Gamma,p}(s) = \prod_{i=0}^{d-1} \zeta_p(s-i)$$

where $\zeta_p(s)$ is the local Riemann zeta function.

(ii) If Γ is the first discrete Heisenberg group (i.e. the two-generator, free nilpotent group of class 2), then

$$\zeta_{\Gamma,p}(s) = \zeta_p(s)\zeta_p(s-1)\frac{(1-p^{-3s+3})}{(1-p^{-2s+2})(1-p^{-2s+3})}.$$

We refer the reader to [16] for further examples. In that paper it is proved that Question 2.13 has a positive answer for the class of finitely generated, torsion-free nilpotent groups. However, using the theory of analytic pro-p goups together with results in logic on quantifier elimination for the analytic theory of the p-adic integers, it is possible to show that Question 2.13 has the following general answer. (An *upper p-chief factor* of Γ is a chief factor of some finite quotient of Γ whose order is divisible by p.)

Theorem 2.15. *Let Γ be a finitely generated group and p a prime. Suppose that: (i) the sequence $a_{p^n}(\Gamma)$ grows at most polynomially; and (ii) the upper p-chief factors of Γ have bounded order. Then $\zeta_{\Gamma,p}(s)$ is rational in p^{-s}.* (See [10] and [11].)

Theorem 2.15 is deduced from the following result, where we write $b_{p^n}(\Gamma)$ for the number of subnormal subgroups of index p^n in Γ:

Theorem 2.16. *Let Γ be a finitely generated group and p a prime. Suppose that $b_{p^n}(\Gamma)$ grows polynomially. Then*

$$\sum_{n=0}^{\infty} b_{p^n}(\Gamma)p^{-ns}$$

is rational in p^{-s}. (See [10] and [11].)

Theorem 2.15 includes all groups of finite upper rank (which are proved in [32] to be the soluble minimax groups). Theorem 2.16 also applies to any subgroup of finite index in $\mathrm{SL}_n(\mathbb{Z})$ for $n \geqslant 3$. It is interesting to compare this with analogous results for Poincaré series associated with the Bass Milnor Wolf invariant. Grunewald has shown that there exist generators for a nilpotent group of Hirsch length 5 with respect to which the corresponding Poincaré series is not rational.

To establish Theorem 2.16 we replace Γ by its pro-p completion $\hat{\Gamma}_p = G$ and observe that $b_{p^n}(\Gamma)$ is the number of open subgroups of index p^n in G (which we previously denoted by $c_{p^n}(G)$). The polynomial growth of this sequence implies by Theorem 2.10 that G is analytic. Theorem 2.16 now follows from:

Theorem 2.17. *Let G be a compact p-adic analytic group. Then*

$$\zeta_G(s) = \sum_{n=0}^{\infty} c_{p^n}(G)p^{-ns}$$

is a rational function in p^{-s}. (See [10] and [11].)

The proof is modelled on the proof, due to Denef and van den Dries, of the conjecture of Serre and Oersterlé that the Poincaré series associated with the p-adic points of a p-adic analytic manifold are rational functions [6]. The methodology involves expressing the Poincaré series as an integral

$$\int_M |f(\mathbf{x})|^s |g(\mathbf{x})||d\nu|$$

where $M \subseteq \mathbb{Z}_p^r$ and $|d\nu|$ is the additive (real-valued) Haar measure on $\mathbb{Z}_p^r$. If M, f and g are *definable* in the language L describing the analytic theory of $\mathbb{Z}_p$, then, using quantifier elimination results for L together with techniques developed by Igusa, Denef and van den Dries are able to evaluate such integrals as rational in p^{-s}. The possibility of expressing $\zeta_G(s)$ as a definable integral of the above form follows from the controlled behaviour of the lower p-series associated with G. We can reduce to the case where G is a uniformly powerful pro-p group. As we saw in Lemma 1.7, G has a global coordinate system $\mathbb{Z}_p^r$, and for each i the ith term G_i of the lower p-series of G can be identified with $p^{i-1}\mathbb{Z}_p^r$. We now associate with each open subgroup $H \leqslant G$ a subset $M(H)$ of $M_r(\mathbb{Z}_p)$ whose rows form coordinates for a basis for the $\mathbb{F}_p$-vector spaces $(H \cap G_i)G_{i+1}/(H^p \cap G_i)G_{i+1}$. We identify definable functions f and g which decode the index of H in G from the measure of the subset $M(H)$. Setting $M = \bigcup_{H \leqslant G} M(H)$ we can then express

$$\zeta_G(s) = \int_M |f(\mathbf{x})|^s |g(\mathbf{x})||d\nu|.$$

We are able to translate filtered group-theoretic statements into statements in the language L describing the analytic theory of the p-adic integers. This translation is used to prove that M, f and g are definable. We refer the reader to [30] for an excellent survey article on the model theory of the p-adic integers.

References

1. H. Bass, 'The degree of polynomial growth of finitely-generated nilpotent groups', Proc. London Math. Soc. (3) **25** (1972), 603-614.
2. H. Bass, M.Lazard and J.-P. Serre, 'Sous-groupes d'indices finis dans $\mathrm{SL}_n(\mathbb{Z})$', Bull. Am. Math. Soc. **70** (1964), 385–392.
3. H. Bass, J. Milnor and J.-P. Serre, 'Solution of the congruence subgroup problem for SL_n $(n \geqslant 3)$ and Sp_{2n} $(n \geqslant 2)$', Inst. Hautes Études Sci. Publ. Math. **33** (1967), 59–137.
4. A. Borel, 'Density and maximality of arithmetic subgroups', J. reine angew. Math. **224** (1966), 78–89.
5. A. Borel and Harish-Chandra, 'Arithmetic subgroups of algebraic groups', Ann. Math. **75** (1962), 485–535.
6. J. Denef and L. van den Dries, 'p-adic and real subanalytic sets', Ann. Math. **128** (1988), 79–138.
7. J.D. Dixon, M. P. F. du Sautoy, A. Mann and D. Segal, 'Analytic pro-p groups', London Math. Soc. Lecture Note Series **157**, Cambridge University Press (1991).
8. M. P. F. du Sautoy, 'Polycyclic groups and topological groups', Supplemento ai Rendiconti del Circolo Matematico di Palermo, **23** (1990), 63–71.
9. M. P. F. du Sautoy, 'Polycyclic groups, analytic groups and algebraic groups', (in preparation).
10. M. P. F. du Sautoy, 'Finitely generated groups, p-adic analytic groups and Poincaré series', Bull. Am. Math. Soc. (1) **23**, (1990) 121–126.
11. M. P. F. du Sautoy, 'Finitely generated groups, p-adic analytic groups and Poincaré series', (in preparation).
12. D. B. A. Epstein, 'Finite presentations of groups and 3-manifolds', Q. J. Math., Oxford, **12** (1961), 205–212.
13. R. Fricke and F. Klein, 'Vorlesungen über die Theorie der elliptischen Modulfunktionen', 2 vols, Teubner, Leipzig (1890, 1892); reprinted by Johnson Reprint, New York (1965).
14. R. I. Grigorchuk, 'On the Hilbert Poincaré series of graded algebras associated with groups', Math. USSR Sbornik **180** (1989), 207–226.
15. M. Gromov, 'Groups of polynomial growth and expanding maps', Inst. Hautes Études Sci. Publ. Math. **53** (1981), 53–78.
16. F. J. Grunewald, D. Segal and G. C. Smith, 'Subgroups of finite index in nilpotent groups', Invent. Math. **93** (1988), 185–223.
17. K. Hensel, 'Über die arithmetischen Eigenschaften der Zahlen', Jahresber. der D. M. V. **16** (1907), 229–319, 388–393, 474–496.

18. J. E. Humphreys, 'Arithmetic groups', in 'Topics in the theory of algebraic groups', Notre Dame Mathematical Lectures Number **10** (1982).

19. G. A. Jones, 'Congruence and non-congruence subgroups of the modular group: a survey', in 'Groups St Andrews 1985', London Math. Soc. Lecture Note Series **121**, Cambridge University Press (1986), 223–234.

20. M. Kneser, 'Strong approximation', Proc. Symp. Pure Math. **9**, Am. Math Soc., Providence R. I. (1966), 187–196.

21. M. Lazard, 'Groupes analytiques p-adiques', Inst. Hautes Études Sci. Publ. Math. **26** (1965), 389–603.

22. A. Lubotzky, 'Combinatorial group theory for pro-p groups', J. Pure Appl. Algebra. **25** (1982), 311–325.

23. A. Lubotzky, 'Group presentation, p-adic analytic groups and lattices in $SL_2(\mathbb{C})$', Ann. Math. **118** (1983), 115–130.

24. A. Lubotzky, 'A group-theoretic characterization of linear groups', J. Algebra **113** (1988), 207–214.

25. A. Lubotzky and A. Mann, 'Powerful p-groups. I: finite groups', J. Algebra **105** (1987), 484–505.

26. A. Lubotzky and A. Mann, 'Powerful p-groups. II: p-adic analytic groups', J. Algebra **105** (1987), 506–515.

27. A. Lubotzky and A. Mann, 'Residually finite groups of finite rank', Math. Proc. Cambridge Philos. Soc. **106** (1989), 385–388.

28. A. Lubotzky and A. Mann, 'On groups of polynomial subgroup growth', (to appear in Invent. Math.).

29. E. Lutz, 'Sur l'équation $y^2 = x^3 - Ax - B$ dans les corps p-adiques', J. Crelle **177** (1937), 237–247.

30. A. Macintyre, 'Twenty years of p-adic model theory', in 'Logic Colloquium 1984', ed. by J. B. Paris, A. J. Wilkie and G. M. Wilmers, North Holland (1986).

31. A. Mann, 'Powerful p-groups', in 'Groups at St. Andrews 1985', London Math. Soc. Lecture Note Series **160**, Cambridge University Press (1991).

32. A. Mann and D. Segal, 'Uniform finiteness conditions in residually finite groups', Proc. London Math. Soc. (3) **61** (1991), 529–545.

33. C.R. Matthews, L.N. Vaserstein and B. Weisfeiler, 'Congruence properties of Zariski-dense subgroups I', Proc. London Math. Soc. **48** (1984), 514–532.

34. J. Milnor, 'A note on curvature and fundamental group', J. Differ. Geom. **2** (1968), 1–7.

35. M. Nori, 'On subgroups of $GL_n(\mathsf{F}_p)$', Invent. Math. **88** (1987), 257–275.

36. V. P. Platonov, 'The problem of strong approximation and the Kneser-Tits conjecture for algebraic groups', Math. USSR-Izv. **3** (1969), 1139–1147.

37. V. P. Platonov, 'Addendum', Math. USSR-Izv. **4** (1970), 784–786.

38. M. S. Raghunathan, 'Discrete Subgroups of Lie Groups', Ergebnisse der Math. und ihrer Grenzgebiete **68**, Springer, Berlin.

39. M. S. Raghunathan, 'On the congruence subgroup problem', Inst. Hautes Études Sci. Publ. Math. **46** (1976), 107–161.

40. M. S. Raghunathan, 'On the congruence subgroup problem, II', Invent. Math. **85** (1986), 73–117.

41. J. G. Ratcliffe, 'Euler characteristics of 3-manifold groups and discrete subgroups of $SL_2(\mathsf{C})$', J. Pure Appl. Algebra **44** (1987), 303–314.

42. D. J. S. Robinson, 'A note on groups of finite rank', Compos. Math. **31** (1969), 240–246.

43. P. Roquette, 'On class field towers', in 'Algebraic Number Theory', ed. by J. W. S. Cassels and A. Frohlich, Academic Press, (1967).

44. D. Segal, 'Subgroups of finite index in soluble groups, I and II', in 'Groups at St Andrews 1985', London Math. Soc. Lecture Note Series **121**, Cambridge University Press (1986), 307–19.

45. D. Segal, 'Residually finite groups', in 'Groups, Canberra 1989', Springer Lecture Notes in Math., **1456** (1991).

46. S. Sen, 'Ramifications in p-adic Lie extensions', Invent. Math. **17** (1972), 44–50.

47. J.-P. Serre, 'Cohomologie galoisienne', Springer Lecture Notes in Math. **5** (1965).

48. J.-P. Serre, 'Le problème des groupes de congruences pour SL_2', Ann. Math. **92** (1970), 489–527.

49. J. Tits, 'Free subgroups in linear groups', J. Algebra, **20** (1972) 250–270

50. A. Weil, 'Sur les fonctions elliptiques p-adiques', C. R. Acad. Sci. **203** (1935), 22.

51. S. Wolf, 'Growth of finitely generated solvable groups and curvature of Riemannian manifolds', J. Differ. Geom., **2** (1968) 421–466.

3.

APPLICATIONS OF THE p-ADIC SUBSPACE THEOREM

G. R. EVEREST

School of Mathematics,
University of East Anglia,
University Plain,
Norwich NR4 7TJ,
England.

Introduction

Given an integer $m \geqslant 2$ and rational numbers $A_1, \ldots, A_m, u_1, \ldots, u_m$ not all zero, we obtain a rational sequence using the linear recurrence relation,

$$u_n = A_1 u_{n-1} + \cdots + A_m u_{n-m}, \quad n > m. \tag{0.1}$$

Sequences of this kind have been studied extensively and the theory has a long and colourful history. One of the earliest and best known is the Fibonacci sequence. This begins $u_1 = u_2 = 1$, and continues via the recurrence relation,

$$u_n = u_{n-1} + u_{n-2}.$$

A slightly more exotic example occurs with the problem of the man who walks along an infinite ladder. Beginning at rung 0 he throws a die with faces numbered 1 to 6, every time moving the number of rungs shown on the upward face of the die. Let u_n denote the probability that he lands on the nth rung. Then a six-term ($m = 6$) recurrence relation is obtained,

$$u_n = 1/6(u_{n-1} + \cdots + u_{n-6}). \tag{0.2}$$

It is amusing to calculate the limiting behaviour of this sequence. In fact,

$$\lim_{n \longrightarrow \infty} u_n = 2/7. \tag{0.3}$$

Although much has been written on this subject (see [10], [15]), it is comparatively recently that strong general results have been obtained about their order of growth. In this article we will present Evertse's theorem from 1984 which gives strong quantitative information about linear recurrence sequences, in *all* valuations. Note that Theorem 1.1 and a weaker form of Theorem 1.2 were proved independently by van der Poorten and Schlickewei (see [11]). The main technical tool is a theorem from Diophantine approximation which generalises much of the research in the earlier half of this century into the approximation of algebraic numbers by rational numbers. Also in this article we will present another application of this tool, which has yielded a new connection with one of the classical invariants in p-adic algebraic number theory: the p-adic regulator of Leopoldt.

1 Linear recurrence sequences

Before Evertse's theorem can be stated, we continue with introductory remarks on linear recurrence sequences (hereafter 'recurrence sequences').

A closed formula for u_n is obtained as follows:

$$u_n = \sum_{j=1}^{r} P_j(n)\alpha_j^n, \tag{1.1}$$

where $\alpha_1, \ldots, \alpha_r$ are algebraic numbers, $\alpha_j \in \overline{\mathbf{Q}}$ and P_j denote polynomials, $P_j[x] \in \overline{\mathbf{Q}}[x]$, with algebraic coefficients. The sequence u_n is called *non-degenerate* if each of the pairwise quotients α_i/α_j, $i \neq j$, is not a root of unity. We study these to save us having to consider sequences like the 'sergeant-major's' sequence $1, 2, 1, 2, 1, 2, \ldots$. In fact a degenerate sequence can be shown to be a finite union of non-degenerate sequences so we have not even lost any generality.

There are several proofs of the formula (1.1) instanced by the following observation. Let

$$\begin{pmatrix} A_1 & \cdots & . & A_m \\ 1 & \cdots & . & 0 \\ 0 & & & 0 \\ \vdots & & & \vdots \\ 0 & \cdots & 1 & 0 \end{pmatrix} \tag{1.2}$$

Then

$$A\mathbf{u}(n) = \mathbf{u}(n+1), \tag{1.3}$$

where $\mathbf{u}(j)$ denotes the vector $(u_{j-1}, \ldots, u_{j-m})^T$. The formula (1.1) follows once a closed expression for the positive integral powers of A is known. But this is readily obtainable by expressing A in Jordan-form. The 'cleanest' case occurs when A is diagonalizable, in which case the polynomials $P_j(x)$ in (1.1) are all constant, and $r = m$. This luxurious assumption, known as the 'simple root' case, will be made in the sequel. Actually the technical details of Evertse's theorem will be kept to a minimum by assuming also that the α_j are all rational. This assumption allows us to get to the heart of the proof quite easily. The paper [4] is recommended as a very clear account of the full technical details.

Suppose that $|\ |$ denotes any valuation of $\mathbf{Q}$. That is, $|\ |$ denotes the ordinary (archimedean) absolute value, or one of the p-adic absolute values $|\ |_p$, where p denotes a (positive) prime number. Here, and throughout, $|\ |_p$ denotes the unique absolute value defined on prime numbers q as follows,

$$|q|_p = \begin{cases} 1 & q \neq p \\ p^{-1} & q = p, \end{cases} \tag{1.4}$$

and extended to $\mathbf{Q}^*$ by multiplicativity. We also take $|0|_p = 0$ as a definition. Write $|\ |_\infty$ for the archimedean absolute value.

Suppose we are in the simple root case and all the P_j and α_j lie in $\mathbb{Q}$. Then it follows that

$$u_n = c\gamma^{-n}\left(\sum_{j=1}^{m} a_j\beta_j^n\right) = c\gamma^{-n}v_n, \tag{1.5}$$

where $c \in \mathbb{Q}, \gamma, a_j, \beta_j \in \mathbb{Z}$ and the β_j are coprime. The growth rate for γ^{-n} in any valuation is obvious. For $|\ |_\infty$ there is nothing to do, v_n will grow like its largest term (or terms). Thus $u_n = 0$ can have only finitely many solutions. It is not so clear what restriction there can be on the p-adic valuations.

Let $\beta = \max\{|\beta_j|_\infty\}$.

Theorem 1.1. *Given v_n as above (see (1.5)). In the rational, simple root case, given $\epsilon > 0$, we have an inequality*

$$\beta^{-n\epsilon} \ll |v_n|_p. \tag{1.6}$$

Putting this another way: let T denote any finite set of finite primes of $\mathbb{Z}$ and $|\ |_{T'}$ the T-primary valuation, i.e. the T-free part. Then

$$\beta^{n(1-\epsilon)} \ll |v_n|_{T'} \ll \beta^n. \tag{1.7}$$

In fact the results of Evertse apply to non-degenerate sequences of algebraic numbers. In the general case it is not even clear what the growth rate at ∞ should be. For example, there could be several complex roots of equal absolute value. How can one guarantee that a large amount of cancellation does not take place?

One of the remarkable features of the results of Evertse for recurrence sequences is that they follow from a much more general point of view. To this day Theorem 1.1 for the case $m > 3$ is accessible only through this level of generality. It should be noted that for cases $m = 2, 3$ there are many better and effective results, obtainable by applying Baker's theorem, together with its multitudinous sharpenings and p-adic analogues (the book [15] is recommended).

Notice that for the P_j, α_j only a finite number of primes appear in the factorizations of the numerator and denominator. Thus, for a sufficiently large set S of prime numbers, $S = \{p_1, \ldots, p_S\}$, we can express u_n as a sum of *S-units*. The term S-unit denotes a rational number composed from primes from S alone, i.e.

$$q \in \mathbb{Q}^* \text{ is an } S\text{-unit means } q = \pm p_1^{e_S} \cdots p_S^{e_S},$$

where $e_i \in \mathbb{Z}, i = 1, \ldots, s$. Evertse's approach is to study a general sum of S-units and show that the absolute value of this sum is about as large as it could be, in any valuation, with some conditions to ensure non-triviality. After all, we have little to say to someone who insists that the equation $x - x = 0$ is a vanishing form of S-units with infinitely many solutions.

Consider $\mathbf{P}^{n-1}(\mathbb{Q})$ the $(n-1)$-dimensional $\mathbb{Q}$-space of all vectors $\mathbf{x} = (x_1, \ldots, x_n)$, where two such are identified if they differ by a non-zero multiple from $\mathbb{Q}$. Given a vector $\mathbf{x} \in \mathbf{P}^{n-1}(\mathbb{Q})$, $\mathbf{x} = (x_1, \ldots, x_n)$, suppose that all the x_i are S-units. Then we may suppose, by clearing denominators, that the x_i are all integers, which have no common factor. Call such a vector *S-admissible*. Finally, given an S-admissible vector $\mathbf{x} \in \mathbb{Z}^n$, write $||\mathbf{x}||$ for some fixed choice of Euclidean norm on $\mathbb{R}^n$. Suppose $S = \{\infty, p_1, \ldots, p_s\}$ is a set of primes.

Theorem 1.2. (See [4], Theorems 1 and 2.)

(i) The equation

$$x_1 + \cdots + x_n = 0, \tag{1.8}$$

has only a finite number of solutions in S-admissible vectors satisfying

$$\sum_{j=1}^{r} x_{i_j} \neq 0 \tag{1.9}$$

for all subsets $\{i_1, \ldots, i_r\} \subseteq \{1, \ldots, n\}$.

(ii) Given $\epsilon > 0, d > 0$ and $T \subseteq S$ a finite subset of S containing ∞. The inequality

$$\prod_{p \in T} |x_1 + \cdots + x_n|_p \ll ||\mathbf{x}||^{1-\epsilon} \tag{1.10}$$

has only a finite number of solutions in S-admissible vectors $\mathbf{x}$ with

$$\sum_{j=1}^{r} x_{i_j} \neq 0 \quad \forall \{i_1, \ldots, i_r\} \subseteq \{1, \ldots, n\}. \tag{1.11}$$

To interpret this, in case 1 obviously we can construct solutions of (1.8) by multiplying the relation $x - x = 0$ by S-units. For example, take $n = 4$ and multiply the equation $x - x = 0$ by different S-units. Adding the results together we obtain an infinite number of projective solutions of the equation $x_1 + x_2 + x_3 + x_4 = 0$. The result says that, fundamentally, this is the only way apart from a finite number of exceptions. For case 2, notice that we do not have strict inclusion on the set of indices of the x_i. Turning the inequality round means an inequality,

$$||\mathbf{x}||^{1-\epsilon} \ll \prod_{p \in T} |x_1 + \cdots + x_n|_p. \tag{1.12}$$

To see the strength of this observe that the upper bound,

$$\prod_{p \in T} |x_1 + \cdots + x_n|_p \ll \|\mathbf{x}\|, \tag{1.13}$$

is obtained trivially.

We will prove Evertse's result in the rational case. His result is more general on two counts. First, he works over an arbitrary algebraic number field. Second, he works with a more general notion of admissibility and this allows him to take care of the polynomial terms in (1.1).

To interpret this theorem for recurrence sequences, part 1 says $u_n = 0$ has only a finite number of solutions provided that we can rule out vanishing subsums. But, as we remarked earlier, the theorem itself says these are generated by relations

$$p_i \alpha_i^n = p_j \alpha_j^n. \tag{1.14}$$

This implies that our sequence is degenerate. Also, part 2 says that in any valuation, provided we take account of obvious divisibility, the sequence is about as large as it could be. This is the statement of Theorem 1.1.

Notice that our man on the ladder is not troubled by any of this. For his sequence, a unique dominant root of 1 is obtained. Thus the sequence u_n (which tends to 2/7) *is* as large as it could be for the infinite valuation. The reader might be interested to consider the growth in the 2- and 3-adic valuations.

2 Diophantine approximation

Suppose α is a real irrational number. It is a classical result of Dirichlet that the inequality

$$|\alpha - p/q| < 1/q^2 \tag{2.1}$$

has infinitely many solutions in pairs of coprime integers p and q, with $q > 0$. To obtain lower bounds, which hold for all coprime pairs p and q, one has to consider special classes of real numbers. Spectacular results have been obtained for the class of real algebraic numbers.

Given the minimal polynomial $f(x)$ of α over $\mathbb{Q}$, with degree d, evaluate at p/q. Then, using the fact that $f(p/q) \neq 0$, one readily obtains the 'Liouville' estimate,

$$1/q^d \ll |\alpha - p/q| . \tag{2.2}$$

Much research has gone into reducing the exponent on the left-hand side. This is essential in solving many classes of Diophantine equations. In 1955, K. Roth gave what is close to best possible when he proved that for $\epsilon > 0$,

$$1/q^{2+\epsilon} \ll |\alpha - p/q| , \tag{2.3}$$

the exponent implicit in the Vinogradov symbol necessarily dependent on ϵ and α. The best possible would be to replace q^ϵ by a power of $\log q$, and a conjecture of Lang (see [7]) asserts this, but the conjecture has resisted proof up to now. See [14] for a proof of Roth's theorem.

For generalizations it pays to express (2.3) in homogeneous form as

$$q^{-1-\epsilon} \ll |q\alpha - p|. \tag{2.4}$$

The equality (2.4) can be generalized to results of the following two types:

(i)

$$||\mathbf{x}||^{-n+1-\epsilon} \ll |\alpha_1 x_1 + \cdots + \alpha_n x_n| \neq 0,$$

where $\alpha_1, \ldots, \alpha_n$ are real algebraic numbers, or

(ii)

$$||\mathbf{x}||^{-1-\epsilon-1/(n-1)} \ll |\alpha_i x_i + x_1|, \quad i = 2, \ldots, n.$$

In fact, it can be shown that these two results are equivalent. Both of these, indeed just about all the (non-effective) results in approximation to algebraic irrationals, follow from the Subspace Theorem of W. Schmidt. This was proved in 1972 and we state it now. Continue to let $||\mathbf{x}||$ denote any fixed choice of Euclidean norm on $\mathbb{R}^n$.

Theorem 2.1. (Subspace Theorem) (See [14], p.153) *Suppose that the linear forms $L_1(\mathbf{x}), \ldots, L_n(\mathbf{x})$ on $\mathbb{R}^n$ are linearly independent with real algebraic coefficients. Given $\epsilon > 0$, the solutions $\mathbf{x} \in \mathbb{Z}^n$ of the inequality,*

$$|L_1(\mathbf{x}) \cdots L_n(\mathbf{x})| \ll \|\mathbf{x}\|^{-\epsilon}, \tag{2.5}$$

lie in a finite number of proper subspaces of $\mathbb{R}^n$.

Notice that one may assume that the subspaces in the statement of the theorem have bases lying in $\mathbb{Q}^n$, for each one may be replaced by the largest subspace generated by the integer solutions.

To see how this implies Roth's theorem, suppose $0 < \alpha < 1$ is real and algebraic. Let $L_1 = q\alpha - p, L_2 = q$, constrained so that $q > p > 0$, and losing no generality in the process. The statement of the Subspace Theorem gives a finite number of one-dimensional subspaces which contain all of the solutions of the inequality,

$$q|\alpha q - p| < q^{-\epsilon}. \tag{2.6}$$

But it is clear that each of the subspaces contain only a finite number of solutions of (2.6) with $p, q \in \mathbb{Z}$, because the left-hand side grows with q while the right-hand side shrinks. Thus in total, (2.6) has only a finite number of integer solutions. Choosing an appropriate constant for $\ll$ gives Roth's theorem (2.3).

The exclusion of complex forms saves technical details. In fact the Subspace Theorem works with these provided that they are chosen in complex conjugate pairs. Such a system of forms is called *symmetric*.

Suppose now p is a (finite) rational prime. One can consider linear forms with coefficients in $\overline{\mathbb{Q}}_p$, the algebraic closure of $\mathbb{Q}_p$. Given a system of linearly independent forms $L_1^{(p)}(\mathbf{x}), \ldots, L_n^{(p)}(\mathbf{x})$ with coefficients in $\overline{\mathbb{Q}}_p$, say that the system is *symmetric* if every form $L_i^{(p)}(\mathbf{x})$ appears as often as its algebraic conjugates over $\mathbb{Q}_p$. Of course this coincides with the definition above when $p = \infty$ and $\mathbb{Q}_p = \mathbb{R}$.

The theorem following was announced by Schlickewei in [12].

Theorem 2.2. p-adic Subspace Theorem *Suppose S denotes a collection of prime numbers which we can take to include the symbol $p_0 = \infty$ (the 'infinite prime', corresponding to the archimedean valuation). Suppose that for each $p \in S$ we have a symmetric system of n linearly independent forms over $\overline{\mathbb{Q}}_p$, say $L_1^{(p)}(\mathbf{x}), \ldots, L_n^{(p)}(\mathbf{x})$. Given $\epsilon > 0$, the solutions $\mathbf{x} \in \mathbb{Z}^n$ of the inequality*

$$\prod_{p \in S} |L_1^{(p)}(\mathbf{x}), \ldots, L_n^{(p)}(\mathbf{x})|_p \ll \|\mathbf{x}\|^{-\epsilon} \tag{2.7}$$

lie inside a finite number of proper subspaces of $\mathbb{Q}^n$.

Notice that the symmetry condition on the linear forms ensures that the product,

$$\prod_{j=1}^{n} L_j^{(p)}(\mathbf{x}),$$

lies in $\mathbb{Q}_p$. In general, of course, one has to consider extensions of the valuations $|\ |_p$.

It is this very general theorem which provides the technical machinery required to prove Theorem 1.2 and consequently the results on recurrence sequences. Now we prove Theorem 1.2 in the special case of rational S-units. The projective nature of the problem means we may consider the following:

$$x_1 + \cdots + x_n, \quad x_i \in \mathbb{Z}, \quad \text{and } S\text{-units}, \tag{2.8}$$

where the x_i have greatest common divisor 1, and

$$\sum_{j=1}^{r} x_{i_j} \neq 0 \quad \forall \{i_1, \ldots, i_r\} \subseteq \{1, \ldots, n\}. \tag{2.9}$$

Proof of Theorem 1.2 The proof goes by induction, the case $n = 1$ being entirely trivial. Suppose the theorem is proved for all integers $0 < n < m$ where $m > 1$, and for all sets S as above, and subsets $T \subseteq S$, with $p_0 \in T$.

The S-unit condition amounts to

$$\prod_{p \in S} \prod_{j=1}^{m} |x_j|_p = 1, \tag{2.10}$$

by the product formula. Consider the S-admissible solutions

$$\mathbf{x} = (x_1, \ldots, x_m) \in \mathbb{Z}^m$$

of the inequality

$$\prod_{p \in T} |x_1 + \cdots + x_m|_p \ll ||\mathbf{x}||^{1-\epsilon}, \tag{2.11}$$

for $T \subseteq S$, $p_0 \in T$. Insert (2.10) into (2.11) to obtain

$$\prod_{p \in T} |x_1 \cdots x_m (x_1 + \cdots + x_m)|_p \prod_{p \in S-T} |x_1 \cdots x_m|_p \ll ||\mathbf{x}||^{1-\epsilon}. \tag{2.12}$$

Notice that one of $x_1, \ldots, x_m$ satisfies

$$||\mathbf{x}|| \ll |x_j|_{p_0} \ll ||\mathbf{x}||. \tag{2.13}$$

Similarly, for each $p \in T$ at least one of $x_1, \ldots, x_m$ satisfies

$$|x_j|_p = 1. \tag{2.14}$$

In order to be able to apply the p-adic Subspace Theorem, we need to omit some of the variables which appear in the product (2.12). For

$p_0 \in T$ omit the variable x_j with (2.13), and cancel $||\mathbf{x}||$ from both sides of (2.12). Note that the constant implicit in (2.13) can be absorbed into that implicit in (2.12). For $p_0 \neq p \in T$ omit the variable x_j which satisfies (2.14), noticing that as p varies it may mean that j varies also.

In other words, for each $p \in T$ we have a collection of linear forms,

$$y_1^{(p)}, \ldots, y_{m-1}^{(p)}, y_1^{(p)} + \cdots + y_{m-1}^{(p)} + y_m^{(p)},$$

where $\{y_1^{(p)}, \ldots, y_m^{(p)}\} = \{x_1, \ldots, x_m\}$. Since they are linear forms in the variable $\mathbf{x}$, label them as

$$\varphi_1^{(p)}(\mathbf{x}), \ldots, \varphi_m^{(p)}(\mathbf{x}).$$

Obviously the $\varphi_i^{(p)}(\mathbf{x})$ are linearly independent on $\mathbf{x} \in \mathbb{R}^m$.

The inequality (2.12) becomes

$$\prod_{p \in T} |\varphi_1^{(p)}(\mathbf{x}) \cdots \varphi_m^{(p)}(\mathbf{x})|_p \prod_{p \in S-T} |x_1 \cdots x_m|_p \ll ||\mathbf{x}||^{-\epsilon}. \tag{2.15}$$

The linear independence of the forms is obvious; so apply the p-adic Subspace Theorem. The conclusion is that a finite number of proper subspaces of $\mathbb{Q}^m$ contain all the solutions of this inequality. Choose one of them, with a basis $B = \{\mathbf{b}_1, \ldots, \mathbf{b}_n\}$, where $n < m$.

The choice of basis allows us to replace the variable x_m by a linear combination of the first $m-1$ variables, the coefficients being dependent upon the choice of B. So let

$$x_m = A_1 x_1 + \cdots + A_{m-1} X_{m-1}.$$

Substitute into the inequality (2.11) to obtain

$$\prod_{p \in T} |B_1 x_1 + \cdots + B_{m-1} x_{m-1}|_p < ||\mathbf{x}||^{-\epsilon}.$$

Cancel any vanishing subsums and clear denominators if necessary. Then we can assume that the coefficients $B_i = 1 + A_i$ are non-zero integers which are coprime. Each of them is an S'-unit for a suitably large finite set S' where $S \subseteq S'$. Write

$$B_i x_i = z_i.$$

Obviously

$$||\mathbf{x}|| \ll ||\mathbf{z}||,$$

and the z_i are S'-units. Thus the vectors $||\mathbf{z}||$ form a collection of S'-admissible vectors which satisfy the inequality

$$\prod_{p \in T} |z_1 + \cdots + z_{m-1}|_p < ||\mathbf{z}||^{1-\epsilon},$$

and

$$\sum_{j=1}^{r} z_{i_j} \neq 0 \quad \forall \{i_1, \ldots, i_r\} \subseteq \{1, \ldots, m-1\}.$$

The induction hypothesis is that only finitely many such vectors exist. Hence the finiteness of the vectors $\mathbf{x}$ is secured.

Repeat this for every one of the finite number of subspaces obtained by applying the p-adic Subspace Theorem. This gives the finiteness statement we require. ◊ ◊

For more recent developments in the theory of sums of S-units consult [5], where there are many generalizations and applications, also a large stock of further references.

Finally in this section, note that the p-adic Subspace Theorem has been generalized recently in a massive way. The new result is known as the Quantitative p-adic Subspace Theorem. The statement is rather complicated, so to give an example application go back to the equation

$$u_n = 0 \tag{2.16}$$

for a general, non-degenerate recurrence sequence. The roots α_j in (1.1) lie in an algebraic extension of degree d, say, over $\mathbb{Q}$. Also, there are at most a finite number of 'primes' dividing the α_j, say s. Then it is now known (see [13]) that the number of solutions of (2.16) is bounded above by a number depending on d and s only. In other words this number does not depend upon the field, only upon the degree over $\mathbb{Q}$.

3 The p-adic regulator

In these final sections we will show how the p-adic Subspace Theorem can be used to count p-primary trace values of the algebraic units in totally real number fields. This will establish a connection with the Leopoldt p-adic regulator. The proof of Theorem 3.1(2) is rather technical so here we present the main thrust of the argument.

Classically one studies the Pell equation,

$$x_1^2 - dx_2^2 = \pm 1, \tag{3.1}$$

for pairs of integers $(x_1, x_2) \in \mathbb{Z}^2$ where $d > 0$ is a square free integer. This is associated naturally with the real quadratic number field $\mathbb{Q}(\sqrt{d})$. By way of generalization, suppose K is a totally real algebraic extension of $\mathbb{Q}$, of degree $r+1$. Let $\mathcal{O}_K$ denote the ring of algebraic integers of K and U_K the group of units of O_K. Choose a $\mathbb{Z}$-basis for $\mathcal{O}_K$, say $\{a_1, \ldots, a_{r+1}\}$. Then the condition that $u \in U_K$ can be expressed as

$$u = a_1 x_1 + \cdots + a_{r+1} x_{r+1} \in U_K$$

if and only if

$$N(\mathbf{x}) = \pm 1. \tag{3.2}$$

Here

$$N(\mathbf{x}) = \prod_{\sigma : K \to \mathbb{R}} |a_1^\sigma x_1 + \cdots + a_{r+1}^\sigma x_{r+1}|,$$

and the product runs over the embeddings of K into $\mathbb{R}$. For example, in the quadratic case we may choose $a_1 = 1, a_2 = \sqrt{d}$ for $d \equiv 2, 3 \bmod 4$, and we recover

$$N(\mathbf{x}) = x_1^2 - dx_2^2, \tag{3.3}$$

as in (3.1). See [8] for more details about algebraic number fields.

Select an element $\alpha \in K$ with the following property. Given p a rational prime, the limit

$$\lim_{u \in U_K} \inf |T(\alpha u)|_p = 0, \tag{3.4}$$

where $T: K \longrightarrow \mathbb{Q}$ denotes the trace map. For example, it may happen in the study of the Pell equation that the coefficient x_2 is infinitely divisible by p. This may be obtained as

$$x_2 = T\left(\frac{u}{2\sqrt{d}}\right), \quad u = x_1 + x_2\sqrt{d} \in U_K.$$

We study the coefficients in the 'norm-form' equation (3.2) via the trace map. More generally we can study the traces of orbits αU_K, and attempt to count them as follows,

$$T(q) = \#\{u \in U_K : |T(\alpha u)| < q\}. \tag{3.5}$$

The structure of U_K is known to be

$$U_K \cong \{\pm 1\} \times \mathbb{Z}^r. \tag{3.6}$$

Choose a basis for $U_K/\{\pm 1\}$, say $\{e_1, \ldots, e_r\}$. Also choose r embeddings $\sigma_i : K \longrightarrow \mathbb{R}, i = 1, \ldots, r$. Define

$$|R_K| = |\det(\log|\sigma_i(e_j)|)|. \tag{3.7}$$

This is known as the *regulator*; it does not depend on the choices made to define it. Write $|n|_{p'}$ for the p-primary part of $n \in \mathbb{Z}$, i.e. $|n|_{p'} = |n|.|n|_p$.

Theorem 3.1.

(i) We have the equation

$$T(q) = \#\{u \in U_K : |T(\alpha u)| < q\} = \theta_1 |R_K|^{-1}(\log q)^r + O((\log q)^{r-1}). \tag{3.8}$$

(ii) For θ_1 a positive rational number, we have the equation

$$\begin{aligned} T_p'(q) &= \#\{u \in U_K : |T(\alpha u)|_{p'} < q\} \\ &= \theta_1 |R_K|^{-1}(\log q)^r + O((\log q)^{r-1} \log\log q). \end{aligned} \tag{3.9}$$

In both cases the formulae can be turned into three-term asymptotic formulae. In fact, the error term in (3.9) turns out to be the correct order of magnitude. Thus the dependence on p in the second formula appears in the second term.

For ease suppose that the prime p splits completely in K. In this case the field K can be embedded in $\mathbb{Q}_p$; for each prime ideal $\mathcal{P}$ in K which divides $p\mathbb{Z}$, the residue degree $(\mathcal{O}_K/\mathcal{P})/(\mathbb{Z}/p\mathbb{Z})$ is 1. Notice that each of the primes $\mathcal{P}$ above p gives rise to a different embedding. Choose r distinct embeddings $\tau_i : K \longrightarrow \mathbb{Q}_p, i = 1, \ldots, r$. Let $\log_p$ denote the p-adic logarithm map defined by

$$\log_p(1 + x) = \sum_{i=1}^{\infty} x^i(-1)^{i+1}/i,$$

which converges for $|x|_p < 1$. Given a generator e_j as above, define $\log_p e_j$ to be $\log_p e_j^{o(e_j)}$, where $o(e_j)$ denotes the order of e_j inside the quotient $\mathbb{Z}_p^*/1 + p\mathbb{Z}_p$. Define *Leopoldt's p-adic regulator* to be

$$|R_{K,p}| = \det(\log_p \tau_i(e_j)), \quad 1 \leqslant i, j \leqslant r. \tag{3.10}$$

This object arose in the study of the p-adic interpolation of the Dedekind zeta function. It is not obvious that $|R_{K,p}|$ should be non-zero but Leopoldt conjectured this is the case. This has been proved in only a few special situations. For example, one of the early applications of p-adic methods in transcendence theory was to prove Leopoldt's conjecture when $K/\mathbf{Q}$ is abelian (see [6] for a very good account of this).

For the remainder of this chapter, we assume the conjecture is true.

4 Proof of Theorem 3.1

We are going to make life easier for ourselves by considering a special case. First, define a subgroup H of U_K, the ***homogeneous hull***, in the following way. Recall (3.10) and put the matrix in Smith normal form. That is, find unimodular matrices R and S over $\mathbb{Z}_p$, with

$$RR_{K,p}S = \begin{bmatrix} p^{e_1} & \dots & \cdots & 0 \\ \vdots & \ddots & & \vdots \\ \vdots & & \ddots & \vdots \\ 0 & \cdots & \cdots & p^{e_r} \end{bmatrix}, \quad e_1 \leqslant \cdots \leqslant e_r.$$

The Leopoldt conjecture asserts that the invariant factors are all non-zero. Reduce modulo p^{e_r+1} and then we can regard R and S as matrices over $\mathbb{Z}/p^{e_r+1}\mathbb{Z}$. Now multiply on the right by the diagonal matrix

$$S' = \begin{bmatrix} p^{e_r-e_1} & \cdots & \cdots & \cdots & \cdots & 0 \\ p^{e_r-e_2} & & & & & \vdots \\ \vdots & & & & & \vdots \\ \vdots & & \ddots & & & \vdots \\ 0 & \cdots & \cdots & \cdots & \cdots & 1 \end{bmatrix}.$$

In this way we can replace the generators $\{e_1, \ldots, e_r\}$ for $U_K/\{\pm 1\}$ by r multiplicatively independent elements $g_1, \ldots, g_r$, such that the matrix

$$H_p = (\log_p \tau_i(g_j))$$

has Smith normal form equal to $\mathrm{diag}(p^e, \ldots, p^e)$, $e_r = e$. Write

$$H = \langle g_1, \ldots, g_r \rangle .$$

Then $H \lhd U_K$ with finite index. Assume also that the g_j satisfy

$$\tau_i(g_j) \equiv 1 \bmod p, \quad 1 \leqslant i, j \leqslant r.$$

In formula (3.9) there is nothing to prove if the values $|T(\alpha u)|_p$ are bounded away from 0, as u runs over U_K. Therefore, make the following assumption:

$$\liminf_{u \in H} |T(\alpha u)|_p = 0. \tag{4.1}$$

The condition can be guaranteed by assuming $\alpha \in \mathcal{O}_K$ has been chosen so that

$$\tau_i(\alpha) \not\equiv 0 \bmod p \quad \text{and} \quad T(\alpha) \equiv 0 \bmod p^e.$$

See Lemma 4.4 for a proof of this. Note in passing that, given H, there will be infinitely many $\alpha \in \mathcal{O}_K$ with this property. This is an easy exercise.

Since H has finite index in U_K, the study of the orbit αU_K comes down to the study of a finite number of orbits αH. If condition (4.1) does not hold then (3.9) follows at once from (3.8). Thus Theorem 3.1 follows from:

Theorem 4.1. *We have the estimates*

(i)

$$\#\{u \in H : |T(\alpha u)| < q\} = A(\log q)^r + O((\log q)^{r-1}),$$

(ii)

$$\#\{u \in H : |T(\alpha u)|_{p'} < q\} = A(\log q)^r + O((\log q)^{r-1} \log\log q),$$

as $q \longrightarrow \infty$.

The basic idea of the proof of Theorem 4.1 is to use the p-adic Subspace Theorem to compare $|T(\alpha u)|$ and $|T(\alpha u)|_{p'}$ with a height function defined on $u \in H$, (see (4.2) and (4.4)). This is defined as follows,

$$\text{for } \lambda \in K,\ h(\lambda) = \max_{\sigma : K \longrightarrow \mathbb{R}} \{\log |\sigma(\lambda)|\}. \tag{4.2}$$

It is an elementary exercise to show that if $u = u(\mathbf{x}) = a_1 x_1 + \cdots + a_{r+1} x_{r+1}$ with respect to the $\mathbb{Z}$-basis of $\mathcal{O}_K$ then $|h(u) - \log \|\mathbf{x}\|\,|$ is bounded independently of u.

Write $t(\lambda) = \log |T(\lambda)|$, with $\lambda \in K$.

The general version of Evertse's theorem in [4] gives:

(i) there are only finitely many $u \in U_K$ for which

$$T(\alpha u) = 0; \tag{4.3}$$

(ii) given $\epsilon > 0$, there is a constant $\theta_2 > 0$, such that

$$\begin{aligned} (1-\epsilon)h(u) - \theta_2 &< t(\alpha u) \\ (1-\epsilon)h(u) - \theta_2 &< t_{p'}(\alpha u) \end{aligned} \tag{4.4}$$

for all $u \in U_K$ with $T(\alpha u) \neq 0$, where $t_{p'}(\alpha u) = \log |T(\alpha u)|_{p'}$.

Write

$$\lambda \in K,\ t_p(\lambda) = \log |T(\lambda)|_p \text{ so that } t_{p'}(\lambda) = t(\lambda) - t_p(\lambda). \tag{4.5}$$

This suggests that the counting of trace values may be effected by counting values of heights. The counting of heights is fairly straightforward but some refinement is needed in order to be able to apply Evertse's theorem.

Note: In view of (4.3), we will always ignore the finite number of elements of U_K with $T(\alpha u) = 0$. This does not affect the kind of results given here.

Lemma 4.2.
We have

$$\#\{u \in H : h(u) < Q\} = AQ^r + O(Q^{r-1}). \tag{4.6}$$

All of the formulae here are capable of considerable improvement. In fact, nearly all of them can be turned into three-term explicit formulae, including an explicit error term, (see [1], [3] and [2]).

Proof Given H with its basis $\{g_1, \ldots, g_r\}$. The functions $\log|\sigma_i(u)|, i = 1, \ldots, r+1$ are linear forms in the exponents of the g_j. Thus we are counting lattice points lying inside a box in $\mathbf{R}^r$ with flat sides. The estimate comes from computing the volume of the box. The leading coefficient can be deduced by an easy integration, being the volume of the box when $Q = 1$. It is

$$A = \theta_3 |R_K|^{-1}, \tag{4.7}$$

where θ_3 is a rational number, depending on r and $[U_K : H]$. ◊ ◊

Before we prove part 1 of Theorem 4.1 observe that 'most' of the units have one conjugate larger than the others. Precisely, given $\theta_4 > 2$, let

$$H_0 = \{u \in H : 0 \neq \log|\sigma_j(u)| - h(u) < -\theta_4\}.$$

Note: Here θ_4 is to be thought of as large and H_0 is then the set of units $u \in H$ with one conjugate much larger than the others. In all that follows, the choice of θ_4 is immaterial and the notation is chosen to honour that fact.

Lemma 4.3.

$$\#\{u \in H_0 : h(u) < Q\} = AQ^r + O(Q^{r-1}). \tag{4.8}$$

In particular

$$\#\{u \in H - H_0 : h(u) < Q\} = O(Q^{r-1}). \tag{4.9}$$

Proof of Lemma 4.3 Again this follows from geometric reasoning: the points which are not in H_0 lie close to hyperplanes in $\mathbf{R}^r$. ◊ ◊

Notice that for elements $u \in H_0$ we have

$$t(\alpha u) = h(u) + O(1). \tag{4.10}$$

Proof of Theorem 4.1(1) Count as follows, letting $Q = \log q$. We have

$$T(Q) = T(q) = \#\{u \in H_0 : t(\alpha u) < Q\} + \#\{u \in H - H_0 : t(\alpha u) < Q\}.$$

The first term is, using (4.10),

$$\#\{u \in H_0 : h(u) + O(1) < Q\}.$$

The presence of the big O term does not affect direct applications of Lemmas 4.2 and 4.3. For the second term use (4.4) so that

$$\begin{aligned}\#\{u \in H - H_0 : t(\alpha u) < Q\} &= O(\#\{u \in H - H_0 : (1-\epsilon)h(u) \\ &< Q + O(1)\}) \\ &= O(\#\{u \in H - H_0 : h(u) = O(Q)\}) \\ &= O(Q^{r-1}) \quad \text{by Lemma 4.3.}\end{aligned}$$

◊ ◊

Proof of Theorem 4.1(2) This time break the counting up as follows:

$$t_{p'}(Q) = \sum_{t=0}^{\infty} \#\{u \in H : t_p(\alpha u) = -t \log p,\ t(\alpha u) < Q + t \log p\}. \quad (4.11)$$

The style of the proof is similar to the above but with several adjustments. For example, the result of Evertse implies that t only runs to a finite limit T (apply (4.4) directly and use the fact that only finitely many $u \in H$ have $h(u)$ bounded by a fixed number). Also one needs to be able to count those units $u \in H$, for which $t_p(u)$ is fixed at $-t \log p$, in a uniform way. This enables the summation over t to do no harm. Rather than give details of this we give an idea of how the p-adic regulator comes into the picture.

The prime p is totally split in K and we have agreed to interpret this via the embeddings $\tau_i : K \longrightarrow \mathbb{Q}_p$, $i = 1, \ldots, r+1$. These determine subgroups H_t of H defined as follows:

$$u \in H_t \Longleftrightarrow \tau_i(u) \equiv 1 \bmod p^t, \quad i = 1, \ldots, r+1. \quad (4.12)$$

The objects of study are solutions $u \in H/H_{t+1}$ of the congruence

$$\alpha_1 u_1 + \cdots + \alpha_{r+1} u_{r+1} \equiv \omega p^t \bmod p^{t+1}, \quad (4.13)$$

where $\omega \in \mathbb{F}_p^*$, $u_i = \tau_i(u)$, $u \in H$, and $\alpha_i = \tau_i(\alpha) \in \mathbb{Z}_p$. It is clear that once a solution has been found, say $u \in H$, then others are obtained as the orbit uH_{t+1}.

Let

$$v_k = u_1 \cdots u_k^2 \cdots u_r, \tag{4.14}$$

so that we study

$$\alpha_1 v_1 + \cdots + \alpha_r v_r + \alpha_{r+1} \equiv \omega' p^t \bmod p^{t+1}. \tag{4.15}$$

The basis $\{g_1, \ldots, g_r\}$ for H gives the following equations:

$$\begin{bmatrix} 2 & 1 & \cdots & 1 \\ 1 & 2 & \cdots & 1 \\ \vdots & \vdots & & \vdots \\ 1 & 1 & \cdots & 2 \end{bmatrix} \begin{bmatrix} \log_p u_1 \\ \log_p u_2 \\ \vdots \\ \log_p u_r \end{bmatrix} = \begin{bmatrix} \log_p v_1 \\ \log_p v_2 \\ \vdots \\ \log_p v_r \end{bmatrix}, \tag{4.16}$$

and

$$\begin{bmatrix} \log_p \tau_1(g_1) & \cdots & \log_p \tau_1(g_r) \\ \vdots & & \vdots \\ \log_p \tau_r(g_1) & \cdots & \log_p \tau_r(g_r) \end{bmatrix} \begin{bmatrix} x_1 \\ \vdots \\ x_r \end{bmatrix} = \begin{bmatrix} \log_p u_1 \\ \vdots \\ \log_p u_r \end{bmatrix}. \tag{4.17}$$

Notice that the matrix with ones and twos is invertible in $M_r(\mathbb{Z}_p)$ for $p > r + 1$. Now put the Leopoldt matrix into Smith normal form. This means that we can find invertible matrices M and N with

$$M \begin{bmatrix} p^e & \cdots & 0 \\ \vdots & \ddots & \vdots \\ 0 & \cdots & p^e \end{bmatrix} N\mathbf{x} = \begin{bmatrix} \log_p v_1 \\ \vdots \\ \log_p v_r \end{bmatrix}, \tag{4.18}$$

with the v_i as in (4.14). Identifying $(\mathbb{Z}/p^{t+1}\mathbb{Z})^r$ with $(\mathbb{Z}_p/p^{t+1}\mathbb{Z}_p)^r$, we thus aim to count solutions $\mathbf{x} \in (\mathbb{Z}_p/p^{t+1}\mathbb{Z}_p)^r$ which satisfy Equation 4.17, with v_i as in (4.14) and (4.16).

The matrix N restricts to a bijection on $(\mathbb{Z}/p^{t+1}\mathbb{Z}_p)^r$ so it is sufficient to count the vectors $\mathbf{y} \in (\mathbb{Z}_p/p^{t+1}\mathbb{Z}_p)^r$ which satisfy

$$M \begin{bmatrix} y_1 p^e \\ \vdots \\ y_r p^e \end{bmatrix} = \begin{bmatrix} \log_p v_1 \\ \vdots \\ \log_p v_r \end{bmatrix}$$

with v_i as in (4.16), and $\mathbf{y} = (y_1, \ldots, y_r)$. Write

$$\begin{bmatrix} p^e y_1 \\ \vdots \\ p^e y_r \end{bmatrix} = M^{-1} \begin{bmatrix} \log_p v_1 \\ \vdots \\ \log_p v_r \end{bmatrix} = \begin{bmatrix} \log_p \omega_1 \\ \vdots \\ \log_p \omega_r \end{bmatrix}. \tag{4.19}$$

It is clear that a solution $\mathbf{y} = (y_1, \ldots, y_r) \in \mathbb{Z}_p^r$ exists if and only if each

$$\omega_i \equiv 1 \bmod p^e.$$

Assume this is the case. Write M (see (4.18)) in the form

$$M = (m_{ij}), \quad 1 \leqslant i, j \leqslant r, m_{ij} \in \mathbb{Z}_p. \tag{4.20}$$

Then the problem comes down to solving the congruence,

$$\alpha_1 \omega_1^{m_{11}} \cdots \omega_r^{m_{1r}} + \cdots + \alpha_r \omega_1^{m_{r1}} \cdots \omega_r^{m_{rr}} + \alpha_{r+1} \equiv \omega' p^t \bmod p^{t+1}, \tag{4.21}$$

where each $\omega_i \equiv 1 \bmod p^e$, $\omega' \in \mathbb{F}_p^*$. Write

$$\omega_i = 1 + p^e z_i. \tag{4.22}$$

Now we are in a position to prove the following.

Lemma 4.4. *Let $N(t)$ denote the number of solutions $\mathbf{x} \in (\mathbb{Z}/p^{t+1}\mathbb{Z})^r$, of the congruence* (4.15), *with* $\mathbf{x}$, $\mathbf{v}$ *as in* (4.16), (4.17). *Then*

$$N(t) = \begin{cases} 0, & t < e \\ (p-1)p^{(t+1-e)(r-1)}, & t \geqslant e. \end{cases}$$

Proof Recall that α was chosen with the properties

$$\begin{aligned} \tau_i(\alpha) &\not\equiv 0 \bmod p \\ T(\alpha) &\equiv 0 \bmod p^e. \end{aligned}$$

Thus, reducing the congruence (4.15) modulo p^e shows there can be no solutions with $t < e$. Therefore suppose $t \geqslant e$.

Now it is clear that for at least one i with $1 \leqslant i \leqslant r$, we must have

$$\alpha_1 m_{1i} + \cdots + \alpha_r m_{ri} \not\equiv 0 \bmod p. \tag{4.23}$$

This congruence cannot fail to be satisfied for all i, otherwise we obtain an equation

$$\begin{bmatrix} m_{11} & \cdots & m_{1r} \\ \vdots & & \vdots \\ m_{r1} & \cdots & m_{rr} \end{bmatrix} \begin{bmatrix} \alpha_1 \\ \vdots \\ \alpha_r \end{bmatrix} = 0.$$

But the matrix M restricts to an invertible matrix over $\mathbb{F}_p$ and $\boldsymbol{\alpha} \not\equiv \mathbf{0} \bmod p$. So, say

$$\alpha_1 m_{11} + \cdots + \alpha_r m_{r1} \not\equiv 0 \bmod p. \tag{4.24}$$

Now the idea is to assign arbitrary values to z_i, $i = 2, \ldots, r$ (see (4.22)), then solve the equations (4.21) and (4.22) for z_1 uniquely.

To justify this last procedure, a method like Hensel's Lemma is required. Once $z_2, \ldots, z_r$ are fixed, we are required to solve a congruence of the form

$$\beta_1(1+p^e z_1)^{m_{11}} + \cdots + \beta_r(1+p^e z_1)^{m_{r1}} + \beta_{r+1} \equiv \omega' p^t \bmod p^{t+1}.$$

We notice that $\alpha_i \equiv \beta_i \bmod p^e$ and $\beta_1 + \cdots + \beta_{r+1} \equiv 0 \bmod p^e$. The condition (4.24) is precisely what is needed to ensure that z_1 is determined uniquely. The coefficients in the p-adic expansion for z_1 are found by induction, in the usual (Hensel) way. Thus $\boldsymbol{\omega}$ determines $\mathbf{v}$ and this determines $\mathbf{x} \bmod (p^{t+1-e}\mathbb{Z})^r$. Also, taking account of $\omega \in \mathbb{F}_p^*$ gives the result stated.

◊ ◊

Finally, it is important to note the following.

Lemma 4.5.

$$H_{t+1} = H^{p^{t+1-e}}, \quad t \geqslant e. \tag{4.25}$$

Proof We simply want to know the solutions of the congruence,

$$(\log_p \tau_i(g_j))\mathbf{x} \equiv 0 \bmod p^{t+1}.$$

Putting the matrix into Smith normal form shows that $\mathbf{x} \in \mathbb{Z}^r$ solves the congruence if and only if $\mathbf{x} \in (p^{t+1-e}\mathbb{Z})^r$. ◊ ◊

Finally we see how these ideas fit together in the proof of Theorem 4.1(2). Recall the formula (4.11),

$$t_{p'}(Q) = \sum_{t=0}^{\infty} \#\{u \in H : t_p(u) = -t\log p,\ t(u) < Q + t\log p\}. \tag{4.26}$$

The same trick as in Theorem 4.1(1) allows us to replace $t(u)$ by $h(u)$. What is harder to deal with is the upper limit for the value of t. Some care must be taken and in fact a diversion is needed (see [2] for the details). A value of the form

$$T = [\log Q / \log p]$$

turns out to be adequate.

Thus (4.26) becomes,

$$t_{p'}(Q) = \sum_{t=0}^{T} \#\{u \in H : t_p(u) - t\log p,\ h(u) < Q + t\log p\}.$$

Now use Lemmas 4.4 and 4.5. The number $N(t)$ in Lemma 4.4 represents the number of solutions, $u \in H/H_{t+1}$, of the equation

$$t_p(u) = -t\log p.$$

We note that $N(t) = 0$ for $t < e$.

We can identify u with its vector of exponents once the basis $\{g_1, \ldots, g_r\}$ of H is fixed. Choose representatives for these solutions, with vectors of exponents $\mathbf{a}_1, \ldots, \mathbf{a}_{N(t)}$. The counting proceeds as follows:

$$t'_p(Q) =$$

$$\sum_{t=0}^{T} \#\{u \in H/H_{t+1},\ v \in H_{t+1} : t_p(u) - t\log p,\ h(uv) < Q + t\log$$

$$= \sum_{t=e}^{T} \sum_{j=1}^{N(t)} \#\{\mathbf{x} \in \mathbb{Z}^r : h(\mathbf{a}_j + p^{t+1-e}\mathbf{x}) < Q + t\log p\},$$

using the fact that $H_{t+1} = H^{p^{t+1-e}}$ (see Lemma 4.5). Thus

$$t'_p(Q) = \sum_{t=e}^{T} \sum_{j=1}^{N(t)} \#\left\{\mathbf{x} \in \mathbb{Z}^r : h(\mathbf{x} + p^{e-t-1}\mathbf{a}_j) < \frac{Q + t\log p}{p^{t+1-e}}\right\}.$$

Now the expression $p^{e-t-1}\mathbf{a}_j$ is a bounded vector so it can be ignored. Expanding out the counting function for h gives a leading term of

$$AQ^r(p-1)\sum_{t=e}^{T} 1/p^{t+1-e}. \tag{4.27}$$

The errors can be kept under control but we give no details here. Notice that the sum in (4.27) differs from the full geometric progression (equal to $(p-1)^{-1}$) by an amount which is

$$O(p^{-T}) = O\left[p^{-\log Q/\log p}\right] = O(Q^{-1}).$$

Thus the leading term comes out to be AQ^r. This should be compared with the leading term obtained by counting just the values of $h(u)$, which can be found at (4.6). This completes the account of the proofs of Theorem 4.1(1) and (2). The reader is referred to [2] for full details ◊ ◊

Final notes

(i) Evertse's result, quoted at (4.4), shows there is a pair of inequalities for $\epsilon > 0$,

$$H(u)^{1-\epsilon} \ll |T(u)|_{p'} \ll H(u).$$

This result as it stands will not give the asymptotic formula in Theorem 4.1(2). However, it does say that if one exists then the leading coefficient should be the same as that for the height counting function.

(ii) We have made heavy use of Leopoldt's conjecture in this work. It would be interesting to work out the effect on the counting procedure if the conjecture were false.

References

1. G. R. Everest, 'Uniform distribution and lattice point counting', to appear. J. Aust. Math. Soc.
2. G. R. Everest, 'p-primary parts of unit traces and the p-adic regulator', to appear.
3. G. R. Everest and J. H. Loxton, 'Counting algebraic units with bounded height', to appear.
4. J.-H. Evertse, 'On sums of S-units and linear recurrences', Compos. Math. **53** (1984) 225–244.
5. J.-H. Evertse, K. Győry, C.L. Stewart and R. Tijdeman, 'S-unit equations and their applications', in Proc. Durham Conference on Transcendence Theory (1986), (ed. A. Baker), 110–174.
6. N. Koblitz, 'p-adic Analysis: a short course on recent work', London Math. Soc. Lecture Notes Series **46**, Cambridge University Press (1980).
7. S. Lang, 'Introduction to Diophantine Approximations', Addison-Wesley, Reading, Massachusetts (1966).
8. S. Lang, 'Algebraic Number Theory', Addison-Wesley, Reading, Massachusetts (1970).
9. H. W. Leopoldt, 'Eine p-adische Theorie der Zetawerte II', J. reine angew. Math. **274/275** (1975), 224–239.
10. A. J. van der Poorten, 'Some facts that should be better known, especially about rational functions', Number Theory and its Applications, (ed. R.A. Mollin), Kluwer Academic (1989).
11. A. J. van der Poorten and H.-P. Schlickewei, 'The growth conditions for recurrence sequences', Macquaire Univ. Reports, 82-0041, North Ryde, Australia (1982).
12. H.-P. Schlickewei, 'The p-adic Thue Siegel Roth Schmidt Theorem', Arch. Math. **29** (1977), 267–270.
13. H.-P. Schlickewei, 'Multiplicities of algebraic linear recurrences', to appear in Acta Mathematica.
14. W. M. Schmidt, 'Diophantine approximation', Springer Lecture Notes in Math. **785** (1980).
15. T. N. Shorey and R. J. Tijdeman, 'Exponential Diophantine Equations', Cambridge University Press (1986).

4.

OUT OF THE p-ADIC INTO THE REAL

MANCHESTER SCHOOL OF p-ADIC ANALYSIS

Mathematics Department,
Manchester University,
Manchester M13 9PL,
England.

Introduction

The theorem of this paper, when applied to the case of a profinite group, is well known, although a proof is not easily located. As a basic result on such groups, we feel that it deserves an explicit statement in the literature. We also give some applications to the continuous representation theory of a profinite group.

We would like to thank the numerous people who made helpful comments on this result.

1 Continuous homomorphisms from profinite groups to CSN groups

We recall that a *profinite group* is a compact Hausdorff topological group in which the family of normal open subgroups forms a fundamental system of neighbourhoods of the identity. The equivalence of this and other definitions is explained in [3], page 7.

Definition 1.1. *A topological group G is a* compact subgroup Noetherian (CSN) group *if and only if it is Hausdorff and any collection of compact subgroups of G has a minimal member with respect to the inclusion ordering.*

It is clear that the class of profinite groups which are also CSN is precisely the class of finite groups. We shall see below that every real or complex Lie group is a CSN group.

Theorem 1.2. *Let G be a profinite group, let H be a CSN group, and let $\varphi: G \longrightarrow H$ be a continuous homomorphism. Then the image of φ is finite.*

Corollary 1.3. *Any continuous homomorphism from a profinite group into a Lie group has finite image.*

Lemma 1.4. *Let $\mathcal{V}$ denote the set of all normal open subgroups of the profinite group G, and let K be a closed subgroup of G. Then*

$$\bigcap_{A \in \mathcal{V}} KA = K.$$

Proof Let $x \in \bigcap_{A\in\mathcal{V}} KA$; then we may write $x = k_A a_A$ for each $A \in \mathcal{V}$, where $k_A \in K$ and $a_A \in A$. The family $\mathcal{V}$ is directed by set inclusion, so that $(k_A)_{A\in\mathcal{V}}$ is a net in K and $(a_A)_{A\in\mathcal{V}}$ is a net in G. But K is compact, so $(k_A)_{A\in\mathcal{V}}$ has a subnet converging to $k \in K$, say. It is then clear that the corresponding subnet of $(a_A)_{A\in\mathcal{V}}$ converges to $k^{-1}x$. But as $\bigcap_{A\in\mathcal{V}} = \{1\}$, any convergent subnet of $(a_A)_{A\in\mathcal{V}}$ must converge to 1, whence $k^{-1}x = 1$, i.e. $x = k$. ◊ ◊

Proof of Theorem 1.2 Keeping our earlier notation, we have that the set $\{\varphi(A) : A \in \mathcal{V}\}$ is a collection of compact subgroups of H, so since H is CSN, there exists a minimal member, i.e. there exists $A_0 \in \mathcal{V}$ such that if $A \in \mathcal{V}$, and $\varphi(A) \subseteq \varphi(A_0)$ then $\varphi(A) = \varphi(A_0)$.

Let $K = \ker\varphi$; then K is a closed subgroup of G. (Note that it is here that we use the Hausdorff property of H.) If $A \in \mathcal{V}$ such that $A \subseteq A_0$, then $\varphi(A) = \varphi(A_0)$ and so $A_0 \subseteq KA$. Hence,

$$A_0 \subseteq \bigcap_{A_0 \supseteq A\in\mathcal{V}} KA = \bigcap_{A\in\mathcal{V}} KA = K$$

by Lemma 1.4.

It follows that G/K is finite, because G/A_0 is finite and so $\varphi(G)$ is finite as required. ◊ ◊

The corollary to Theorem 1.2 now follows at once from the following proposition.

Proposition 1.5. *Any Lie group is a CSN group.*

Proof Let $\mathcal{C}$ be a family of compact subgroups of a Lie group H. Let $\mathcal{C}_1$ be the subfamily containing all $B \in \mathcal{C}$ for which $\dim B$ is minimal. Now let $\mathcal{C}_2$ be the subfamily of $\mathcal{C}_1$ containing all B for which $|B/B^0|$ is minimal (over $\mathcal{C}_1$). Then if $B_0 \in \mathcal{C}_2$ and $B \in \mathcal{C}$ such that $B \subseteq B_0$, we have $B^0 \subseteq B_0^0$, so as $B_0 \in \mathcal{C}_1$, we have $\dim B_0 = \dim B_0^0$, giving $B^0 = B_0^0$. Clearly, $B \in \mathcal{C}_1$, so as $B_0 \in \mathcal{C}_2$, and

$$B/B^0 = B/B_0^0 \subseteq B_0/B_0^0,$$

we have $B/B^0 = B_0/B_0^0$. Hence, $B = B_0$ and B_0 is minimal in $\mathcal{C}$. ◊ ◊

2 Some applications

One of the main source of applications of Theorem 1.2 is in the continuous representation theory of a profinite group. We describe some examples in this area. Throughout this section let G be a profinite group. We restrict attention to complex representations although similar results hold over the real numbers $\mathbf{R}$ and even the quaternions $\mathbf{H}$.

Let σ be a continuous representation of G on the finite-dimensional complex vector space V. Then by Corollary 1.3, σ factors through a finite quotient group

$$\sigma: G \longrightarrow G/L_0 \longrightarrow \mathrm{GL}(V)$$

where L_0 is a compact open normal subgroup of G, and we then have

$$\sigma: G \longrightarrow G/L_0 \longrightarrow \mathrm{U}(V) \longrightarrow \mathrm{GL}(V)$$

where $\mathrm{U}(V)$ is the unitary group of V with respect to some hermitian inner product with respect to which σ is unitary. Hence we have the following results.

Let $\mathrm{Rep}^{\mathrm{c}}_{\mathbf{C}}(G)$ denote the continuous representation ring of G, and for a finite group H let $\mathrm{Rep}_{\mathbf{C}}(H)$ denote the ordinary representation ring of H. Similarly, let $\mathrm{Irred}^{\mathrm{c}}_{\mathbf{C}}(G)$ denote the set of irreducible continuous characters of G and let $\mathrm{Irred}_{\mathbf{C}}(H)$ denote the set of irreducible characters of H.

Proposition 2.1. *We have*

$$\begin{aligned} \mathrm{Rep}^{\mathrm{c}}_{\mathbf{C}}(G) &\cong \lim_{\longrightarrow} \mathrm{Rep}_{\mathbf{C}}(G/L), \\ \mathrm{Irred}^{\mathrm{c}}_{\mathbf{C}}(G) &\cong \lim_{\longrightarrow} \mathrm{Irred}_{\mathbf{C}}(G/L), \end{aligned}$$

where L ranges over the directed system of all compact open normal subgroups of G.

In particular, we can take the case of $G = \mathrm{GL}_N(\mathbf{Z}_p)$. Then G admits the finite quotients $\mathrm{GL}_N(\mathbf{Z}/p^n\mathbf{Z})$ where $n \geqslant 1$. The case $n = 1$ yields a finite group of Lie type $\mathrm{GL}_N(\mathbf{Z}/p\mathbf{Z})$; it is worth noting that if p is odd, then any subgroup $H \leqslant \mathrm{GL}_N(\mathbf{Z}_p)$ of order prime to p is mapped isomorphically onto a subgroup of $\mathrm{GL}_N(\mathbf{Z}/p\mathbf{Z})$, and if $p = 2$, the same is true if we map into the quotient $\mathrm{GL}_N(\mathbf{Z}/4\mathbf{Z})$. The complex representation theory of $\mathrm{GL}_2(\mathbf{Z}/p^n\mathbf{Z})$ forms the subject of [2] which also contains conjectures for the complex representation theory of $\mathrm{GL}_N(\mathbf{Z}/p^n\mathbf{Z})$ in general. For more recent work on complex representations of finite groups of Lie type see [1].

References

1. F. Digne and J. Michel, 'Representations of finite groups of Lie type', London Math. Soc. Student Text **21** (1991).

2. G. Lusztig, 'Representations of GL_2', Lecture Notes, Warwick University (1973).

3. S. S. Shatz, 'Profinite Groups, Arithmetic and Geometry', Ann. of Math. Studies **67** (1972).

5.

COUPLING CONSTANTS FOR p-ADIC GROUPS

R. J. PLYMEN

Mathematics Department,
Manchester University,
Manchester M13 9PL,
England.
<mbbgsrp@uk.ac.mcc.cms>

Introduction

The work of Vaughan Jones has brought into prominence the theory of II_1 factors, their traces and coupling constants. Factors of type II_1 were discovered by Murray and von Neumann. They are weakly closed, star algebras of operators on a Hilbert space, with centre consisting of the scalars, and equipped with a finite, positive, trace functional. Such a factor N provides a scale by which to measure the relative dimension of the Hilbert space H on which it acts. We write this number as $\dim_N(H)$. In this case there are no restrictions on the number. Indeed it may vary continuously from 0 to ∞. Jones varied this idea and looked at a pair of factors $N \subseteq M$ considering the dimension of M relative to its subfactor N. This he defined as $[M : N] = \dim_N(H)/\dim_M(H)$, showing it to be independent of the Hilbert space. The importance of this idea became evident in the restrictions found on the index. The allowable set of indices is precisely $\{4\cos^2(\pi/n) : n \geqslant 3\}$ and the interval $[4, \infty]$.

Elsewhere in mathematics, the modern theory of automorphic forms as envisaged by Langlands gives equal weight to the representation theory of p-adic reductive groups and the representation theory of real groups: see the companion article by Shahidi in this book.

At first glance, the world of coupling constants is far removed from the world of representation theory of p-adic groups. However, we shall show by means of a single theorem that a bridge can be built between these two areas. For the sake of concreteness, we state the result for the classical family of simple groups $\mathrm{PSL}_n(F)$. Here F is a local field, $\mathrm{SL}_n(F)$ is the special linear group of all $n \times n$ matrices over F with determinant 1, and $\mathrm{PSL}_n(F)$ is the quotient of $\mathrm{SL}_n(F)$ by its centre. The field F is a non-discrete locally compact field, and $\mathrm{PSL}_n(F)$ thereby acquires the structure of a locally compact group. We recall that an irreducible unitary representation π of $\mathrm{PSL}_n(F)$ is in the *discrete series* if matrix coefficients of π are square integrable. Also a discrete subgroup of $\mathrm{PSL}_n(F)$ with finite covolume is called a *lattice*.

Theorem 0.1. *Let Γ be a lattice in $\mathrm{PSL}_n(F)$. Let π be a representation of $\mathrm{PSL}_n(F)$ on the Hilbert space H which is in the discrete series. Let M be the von Neumann algebra on H generated by $\pi(\Gamma)$. Then M is a II_1 factor and*

$$\dim_M(H) = \operatorname{covol}(\Gamma) \cdot d_\pi$$

where $\operatorname{covol}(\Gamma)$ *is the covolume of* Γ *and* d_π *is the formal degree of* π.

This is an exact analogue for p-adic groups of an established result for real Lie groups [4], 3.3.2, which in turn is essentially (3.3) in Atiyah and Schmid [1]. In the course of our proof, we shall follow [4], p.142–147, closely. The definitions of coupling constant, formal degree and covolume are such as to make our proof self-contained, modulo an important theorem of Wang [8] on density properties of finite covolume discrete subgroups of locally compact groups.

The group $\mathrm{PSL}_n(F)$ has a plentiful supply of discrete series representations. But discrete subgroups are not so easy to find. The subgroup $\mathrm{PSL}_n(\mathbb{Z})$ of $\mathrm{PSL}_n(\mathbb{Q}_p)$ is, of course, not discrete; it is relatively compact, its closure being $\mathrm{PSL}_n(\mathbb{Z}_p)$. We can, however, proceed as follows. Let $\mathbb{F}_q$ be a finite field and let $F = \mathbb{F}_q((1/x))$ be the Laurent series field with valuation at infinity given by $\operatorname{val}(1/x) = 1$. Then F is a local field of characteristic q. It comprises formal Laurent series $\sum_{n \leqslant N} \alpha_n x^n$ with $\alpha_n \in \mathbb{F}_q$. Infinitely many negative powers n are allowed, finitely many positive powers, for the norm of $1/x^n$ is given by

$$\left\| \frac{1}{x^n} \right\| = \left\| \frac{1}{x} \right\| \cdots \left\| \frac{1}{x} \right\| = q^{-\operatorname{val}\left(\frac{1}{x}\right)} \cdots q^{-\operatorname{val}\left(\frac{1}{x}\right)} = q^{-n}.$$

Let $\Gamma = \mathrm{PSL}_n(\mathbb{F}_q[x])$ where $\mathbb{F}_q[x]$ is the polynomial ring over $\mathbb{F}_q$ in one variable x. Since the valuation is a homomorphism $\operatorname{val}: F^\times \longrightarrow \mathbb{Z}$ it follows that $\operatorname{val}(x) = -\operatorname{val}(1/x) = -1$ so that

$$\|x\| = q^{-\operatorname{val}(x)} = q.$$

Hence $\mathrm{PSL}_n(\mathbb{F}_q[x])$ is a discrete subgroup of $\mathrm{PSL}_n(\mathbb{F}_q((1/x)))$.

Let $\mathcal{O} = \mathbb{F}_q[[1/x]]$ denote the compact ring of power series in $1/x$. We now set $n = 2$. Then $K = \mathrm{PSL}_2(\mathcal{O})$ is a maximal compact subgroup of

$G = \mathrm{PSL}_2(F)$. Following Serre [7], p.82, we put on G the unique Haar measure μ such that

$$\mu(K) = q - 1.$$

Now K is a profinite group which maps onto the finite group $\mathrm{PSL}_2(\mathbf{Z}/q\mathbf{Z})$. Let τ be a cuspidal representation of the finite group $\mathrm{PSL}_2(\mathbf{Z}/q\mathbf{Z})$. This means that τ is not a component in the principal series of $\mathrm{PSL}_2(\mathbf{Z}/q\mathbf{Z})$. Now $\mathrm{PSL}_2(\mathbf{Z}/q\mathbf{Z})$, at least when q is odd, admits cuspidal representations of degree $q-1$ (see [3], p.233). Let τ be such a representation

$$\tau \colon \mathrm{PSL}_2(\mathbf{Z}/q\mathbf{Z}) \longrightarrow \mathrm{GL}_{q-1}(\mathbf{C})$$

and let σ be the composite

$$\sigma \colon \mathrm{PSL}_2(\mathcal{O}) \longrightarrow \mathrm{PSL}_2(\mathbf{Z}/q\mathbf{Z}) \longrightarrow \mathrm{GL}_{q-1}(\mathbf{C}).$$

Then, in the spirit of Kutzko's address at the ICM 1986 [5], the induced representation $\pi = \mathrm{Ind}_K^G(\sigma)$ will be a supercuspidal representation of G with formal degree given by

$$d_\pi = d(\sigma)/\mathrm{vol}\,(K) = 1$$

where $d(\sigma)$ is the degree of σ.

Now the covolume of Γ is given by [7], p.89,

$$\mathrm{covol}\,(\Gamma) = 1/(q^2 - 1)$$

so we have

$$\dim_M(H) = \mathrm{covol}\,(\Gamma) \cdot d_\pi = \frac{1}{q^2 - 1}.$$

1 Lattices in semisimple groups over a local field

Now let F be a local field with finite residue field. The valuation on F will be denoted val. The norm of F is given by $|x|_F = q^{-\mathrm{val}(x)}$ where q is the cardinal of the residue field.

An algebraic group $\mathbf{H}$ defined over F is *F-isotropic* if $\mathbf{H}$ contains an F-split torus.

From now on, $\mathbf{G}$ will be a connected semisimple linear algebraic group defined over F such that each factor $\mathbf{G}_i$ is a connected simple F-isotropic subgroup. Let G be the subgroup of F-rational points of $\mathbf{G}$. We suppose in addition that G has trivial centre. The most prominent example is the classical group $\mathrm{PSL}_n(F)$ which we chose to use in the Introduction. This is a simple group of Lie type [2].

The group G has two distinct topologies. Topology (i) comes from the non-discrete locally compact topology of F; in this topology, each connected component of G is a point. Topology (ii) is the Zariski topology; in this topology the whole of G is a single connected component. When we refer to Γ as a discrete subgroup of G, we are using topology (i).

Let μ denote a chosen Haar measure on G. Let Γ be a discrete subgroup of G. A fundamental domain for Γ in G is a subset D of G which is measurable and satisfies the following condition.

- The sets $\gamma_1 D \bigcap \gamma_2 D$ (for $\gamma_1, \gamma_2 \in \Gamma$ with $\gamma_1 \neq \gamma_2$) and $G - \bigcup_{\gamma\in\Gamma} \gamma D$ have null measure.

The Haar measure $\mu(D)$ of D is the *covolume* of Γ in G. A finite covolume discrete subgroup is called a *lattice.*

Definition A discrete group Γ is an *icc group (infinite conjugacy class group)* if each conjugacy class except one is infinite.

The following result came out of discussions with M. McCrudden.

Lemma 1.1. *Let Γ be a Zariski dense subgroup of a Zariski connected centreless group G. Then Γ is icc.*

Proof Let $h \in \Gamma$ and let $C(h,\Gamma)$ be its conjugacy class in Γ. The map $\Gamma \to C(h,\Gamma), \gamma \longmapsto \gamma h \gamma^{-1}$ extends in topology (ii) to a continuous map $\varphi: \overline{\Gamma} \longrightarrow \overline{C(h,\Gamma)}$ between Zariski closures.

If $C(h,\Gamma)$ is finite then $C(h,\Gamma) = \overline{C(h,\Gamma)}$ since 'points are closed' in the Zariski topology. Since $\overline{\Gamma} = G$, we have $\varphi: G \to C(h,\Gamma)$, $g \longmapsto ghg^{-1}$; but by definition of conjugacy class, this map has image equal to $C(h,G) \supseteq C(h,\Gamma)$ and hence is surjective. Now the pre-image of h is the centralizer $Z(h,G)$. The centralizer $Z(h,G)$ is a closed subgroup of finite index in G. But G is Zariski connected and so $Z(h,G) = G$; otherwise the cosets of

$Z(h, G)$ would partition G into finitely many open sets. Hence h lies in the centre $Z(G)$. But G has trivial centre, so $h = e$. It follows that if $h \neq e$ then $C(h, \Gamma)$ is infinite. $\Diamond\Diamond$

Lemma 1.2. *Let Γ be a lattice in G. Then Γ is an icc group.*

Proof The lattice Γ is Zariski dense in G by Wang's density theorem [8], (1.4); see also Prasad [6], (1.12). The result now follows from Lemma 1.1. $\Diamond\Diamond$

2 Factors of type II_1

We will prove the following theorem.

Theorem 2.1. *If Γ is a lattice in G, (π, H) is in the discrete series of G, M is the II_1 factor generated by $\pi(\Gamma)$, then the coupling constant of the M-module H is given by*

$$\dim_M(H) = \operatorname{covol}(\Gamma) \cdot d_\pi.$$

Proof Let λ be the left regular representation of Γ on the Hilbert space $\ell^2(\Gamma)$, and let $vN(\Gamma)$ be the von Neumann algebra generated by the image of Γ, so that $vN(\Gamma)$ is the bicommutant $\lambda(\Gamma)''$. We shall write

$$M = vN(\Gamma) = \lambda(\Gamma)''.$$

Now Γ is an icc group by Lemma 1.2; hence M is a II_1 factor. The normalized trace is specified by

$$\operatorname{tr}_M\left(\sum \alpha_\gamma \lambda(\gamma)\right) = \alpha_e.$$

Form the Hilbert space $L^2(G)$ with respect to Haar measure μ. Let L be the left regular representation of G on $L^2(G)$. Let $p \in B(L^2(G))$ be a projection in the commutant $L(G)'$ which defines an irreducible representation of G. Denote by H_p the range of p, by $\pi_p\colon \Gamma \to U(H_p)$ the corresponding subrepresentation of $L \mid \Gamma$, and by $\pi_p(\Gamma)''$ the von Neumann algebra generated by $\pi_p(\Gamma)$ in $B(H_p)$. Consider the morphism

$$L(\Gamma)'' \to \pi_p(\Gamma)'', x \longmapsto pxp.$$

Now H_p and its orthogonal complement $H_p^\perp$ are stable under all operators in $L(\Gamma)$. Hence the von Neumann algebra $L(\Gamma)''$ is the direct sum of two von Neumann algebras, one acting on H_p and the other acting on $H_p^\perp$. So the morphism is surjective.

The isomorphism $L^2(G) \to \ell^2(\Gamma) \otimes L^2(D)$ which sends φ to $\sum \delta_\gamma \otimes \varphi_\gamma$ where $\delta_\gamma \in \ell^2(\Gamma)$ is the characteristic function of $\{\gamma\}$ in Γ, and where $\varphi_\gamma(g) = \varphi(\gamma g)$ for $\gamma \in \Gamma$, $g \in D$, has the property that $L \mid \Gamma = \lambda \otimes 1$ on $\ell^2(\Gamma) \otimes L^2(D)$. Hence

$$L(\Gamma)'' = (\lambda(\Gamma) \otimes 1)'' = vN(\Gamma) \otimes \mathbf{C} \cong vN(\Gamma).$$

But $vN(\Gamma)$ is a simple ring, and so the morphism is injective. So we have an isomorphism

$$L(\Gamma)'' \cong M.$$

We are given a representation $\pi\colon G \to U(H)$ in the discrete series of G. By the previous two paragraphs we may assume that H is included as an M-module in $L^2(G)$. This inclusion, say u, satisfies $u^*u = id_H$ and $uu^* = p$ where p is the projection from $L^2(G)$ onto H.

Now the map

$$M \to \ell^2(\Gamma), x \longmapsto x(\delta_e)$$

is a linear injection which secures a Hilbert-space isomorphism $L^2(M) \to \ell^2(\Gamma)$. We recall that $L^2(M)$ is the completion of M in the $\|\ \|_2$ norm given by $\|x\|_2 = (\mathrm{tr}_M(xx^*))^{\frac{1}{2}}$. The map $L^2(M) \to \ell^2(\Gamma)$ is an M-module map, where $L^2(M)$ is viewed as a M-module by left multiplication. Then the M-module $L^2(G)$ may be identified with $L^2(M) \otimes K$ where K is the trivial M-module $L^2(D)$ associated to some fundamental domain D of Γ in G.

We have the inclusion of M-modules

$$H \xrightarrow{u} L^2(M) \otimes K.$$

Under these circumstances, the coupling constant $\dim_M(H)$ is defined to be

$$\dim_M(H) = \mathrm{Tr}_{M'}(u^*u)$$

where M' is the commutant $\mathrm{End}_M(L^2(M) \otimes K)$ and $\mathrm{Tr}_{M'}$ is the natural trace on M'. This commutant M' is generated by finite sums

$$x = \sum \rho_\gamma \otimes a_\gamma$$

where

$$\rho_\gamma = J\lambda(\gamma)J \in \mathrm{End}_M L^2(M)$$

and a_γ is a finite rank operator in $B(K)$. Here, J is the conjugation on $L^2(M)$ determined by $Jx = x^*$ on the dense subspace M. Now

$$\mathrm{Tr}_{M'}(\rho_\gamma \otimes a_\gamma) = \mathrm{tr}_M \lambda(\gamma).T_K(a_\gamma)$$

where tr_M is the normalized trace on M, and T_K is the ordinary trace defined on trace-class operators in $B(K)$. This equation holds *by definition* of the natural trace $\mathrm{Tr}_{M'}$. But $\mathrm{tr}_M \lambda(\gamma) = 0$ unless $\gamma = e$. Hence

$$\mathrm{Tr}_{M'}(x) = T_K(a_e).$$

Let $q : L^2(G) \to K$ be the projection sending φ to $\varphi \mid D$, and let T be the ordinary trace on trace-class operators in $B(L^2(G))$. We have

$$L^2(G) = L^2(M) \otimes K = \ell^2(\Gamma) \otimes K.$$

Let $x = \sum \rho_\gamma \otimes a_\gamma$ as before, and consider the compression qxq of x to K. The space K is viewed as a subspace of $\ell^2(\Gamma) \otimes K$ by the embedding

$\xi \longmapsto \delta_e \otimes \xi$. Now the operators $\{\rho_\gamma\}$ permute the basis vectors $\{\delta_\gamma\}$ amongst themselves. Only the identity ρ_e on $\ell^2(\Gamma)$ will fix δ_e. Hence

$$T(qxq) = T(q \cdot I \otimes a_e \cdot q) = T_K(a_e).$$

So we have

$$\mathrm{Tr}_{M'}(x) = T(qxq).$$

This will hold, by continuity, for the projection p from $L^2(G)$ onto H:

$$\dim_M(H) = \mathrm{Tr}_{M'}(p) = T(qpq).$$

Let (ε_n) be an orthonormal base in K, so that $q\varepsilon_n = \varepsilon_n$. Then

$$T(qpq) = \sum_n \langle qpq\varepsilon_n \mid \varepsilon_n\rangle = \sum \langle p\varepsilon_n \mid \varepsilon_n\rangle = \sum \|p\varepsilon_n\|^2.$$

Hence

$$\dim_M(H) = \sum \|p\varepsilon_n\|^2.$$

The orthonormal base $(\delta_\gamma \otimes \varepsilon_n)$ of $\ell^2(\Gamma) \otimes K$ is also the base $(L(\gamma)\varepsilon_n)$ of $L^2(G)$. Let now η be a unit vector in $H \subseteq L^2(G)$ so that $p\eta = \eta$. By the Parseval identity:

$$1 = \|L(g)\eta\|^2 = \sum_\gamma \sum_n |\langle L(g)\eta \mid L(\gamma)\varepsilon_n\rangle|^2.$$

Then

$$\begin{aligned} \mathrm{covol}(\Gamma) &= \int_D dg = \sum_n \sum_\gamma \int_D |\langle L(\gamma^{-1}g)p\eta \mid \varepsilon_n\rangle|^2 dg \\ &= \sum_n \int_G |\langle pL(g)\eta \mid \varepsilon_n\rangle|^2\, dg \end{aligned}$$

since $p \in L(G)'$. We also shifted variable so as to integrate over all translates γD of the fundamental domain D. We next have

$$\mathrm{covol}(\Gamma) = \sum_n \int_G |\langle L(g)\eta \mid p\varepsilon_n\rangle|^2\, dg.$$

The integrand is the square of the modulus of the matrix coefficient

$$g \longmapsto \langle L(g)\eta \mid p\varepsilon_n\rangle$$

taken with respect to the vectors η, $p\varepsilon_n$ both in H. But this is the matrix coefficient of the representation π in the discrete series of G. By the Schur relation we have

$$\int_G |\langle L(g)\eta \mid p\varepsilon_n\rangle|^2 \, dg = d_\pi^{-1}\|\eta\|^2\|p\varepsilon_n\|^2$$

where d_π is *by definition* the formal degree of π. Finally we have

$$\operatorname{covol}(\Gamma) = \sum_n d_\pi^{-1}\|p\varepsilon_n\|^2 = d_\pi^{-1} \dim_M(H).$$

This completes the proof of the theorem. ◊ ◊

References

1. M. F. Atiyah and W. Schmid, 'A geometric construction of the discrete series for semisimple Lie groups', Invent. Math. **42** (1977), 1–62.

2. R. W. Carter, 'Simple groups of Lie type', Wiley Classics Library (1989).

3. L. Dornhoff, 'Group representation theory, part A', Marcel Dekker, New York (1971).

4. F. M. Goodman, P. de la Harpe, and V. F. R. Jones, 'Coxeter graphs and towers of algebras', MSRI publication 14, Springer-Verlag (1989).

5. P. C. Kutzko, 'On the supercuspidal representations of GL_N and other p-adic groups', Proceedings ICM (1986), 853–861.

6. G. Prasad, 'Lattices in semisimple groups over local fields', Studies in algebra and number theory, (ed. by G.-C. Rota), Academic Press, New York (1979), 285–356.

7. J.-P. Serre, 'Trees', Springer-Verlag, Berlin (1980).

8. S. P. Wang, 'On density properties of S-subgroups of locally compact groups', Ann. Math. **94** (1971), 325–329.

6.

THE LOCAL FERMAT PROBLEM

PAULO RIBENBOIM

Department of Mathematics,
Queen's University,
Kingston,
Ontario K7L 3N6,
Canada.

Introduction

Our aim is to investigate the solution of Fermat's equation in the local fields $\mathbb{Q}_p$ of p-adic numbers. In order to have a mostly self-contained presentation, we shall begin with preliminaries about the resultant and Hensel's lemma. First, we recall some notation and well-known facts.

Let $\mathbb{Q}_p$ be the field of p-adic numbers and $\mathbb{Z}_p$ the ring of p-adic integers. We denote by ν_p the p-adic valuation; the values are integers, the maximal ideal consists of the multiples of p and the residue field is $\mathbb{Z}_p/p\mathbb{Z}_p \cong \mathbb{F}_p$ (the field with p elements).

For every $a \in \mathbb{Z}_p$ we denote by $\bar{a}$ the residue class $a \bmod p$; so $\bar{a} \in \mathbb{F}_p$. We note that $\mathbb{Z}_p \cap \mathbb{Q} = \mathbb{Z}$.

If $a, b \in \mathbb{Q}_p$ and $n \geqslant 1$, we write $a \equiv b \bmod p^n$ whenever p^n divides $a - b$, that is there exists $c \in \mathbb{Z}_p$ such that $a - b = cp^n$.

Any element $a \in \mathbb{Z}_p$ such that $\nu_p(a) = 0$ is called a unit.

$\mathbb{Q}_p$ is a complete metric topological field with respect to the p-adic distance $d_p(a, b) = e^{-\nu_p(a-b)}$.

If $f(X) = a_0X^n + a_1X^{n-1} + \cdots + a_n \in \mathbb{Q}_p[X]$, we make the definition $\tilde{\nu}_p(f) = \min_{0 \leqslant i \leqslant n}\{\nu_p(a_i)\}$. If $f, g \in \mathbb{Q}_p[X]$, with $g \neq 0$, we define

$$\tilde{\nu}_p\left(\frac{f}{g}\right) = \tilde{\nu}_p(f) - \tilde{\nu}_p(g),$$

which is well-defined. Then $\tilde{\nu}_p$ is a valuation of the field $\mathbb{Q}_p(X)$, whose restriction to $\mathbb{Q}_p$ is the valuation ν_p. For simplicity, we shall write ν_p, instead of $\tilde{\nu}_p$.

If $f, g \in \mathbb{Q}_p[X]$, we write $f \equiv g \bmod p^n$ when $\nu_p(f-g) \geqslant n$, or equivalently, p^n divides each coefficient of $f - g$.

For every $f = \sum_{i=0}^{m} a_i X^i \in \mathbb{Z}_p[X]$, we denote by $\bar{f} = f \bmod p$ the polynomial $\sum_{i=0}^{m} \overline{a_i} X^i \in \mathbb{F}_p[X]$.

We recall now some well-known facts about polynomials in $\mathbb{Q}_p[X]$.

The polynomial $f \in \mathbb{Z}_p[X]$ is said to be *primitive* when $\nu_p(f) = 0$. Every polynomial $f \in \mathbb{Z}_p[X]$ may be written as $f = af_1$, where $a \in \mathbb{Z}$, $f_1 \in \mathbb{Z}_p[X]$ and f_1 is primitive.

0.1. Gauss's Lemma: *If $f, g \in \mathbb{Z}_p[X]$ are primitive polynomials, then $f \cdot g$ is also a primitive polynomial.*

0.2. *If $f \in \mathbb{Z}_p[X]$ is primitive and $f = g \cdot h$ with $g, h \in \mathbb{Q}_p[X]$, then $f = f_1 \cdot g_1$ for some primitive polynomials $f_1, g_1 \in \mathbb{Z}_p[X]$, such that* $\deg(g_1) = \deg(g)$, $\deg(h_1) = \deg(h)$.

The non-constant polynomial $f \in \mathbb{Z}_p[X]$ (resp. $f \in \mathbb{Q}_p[X]$) is *irreducible* in $\mathbb{Z}_p[X]$ (resp. in $\mathbb{Q}_p[X]$) if it is impossible to write $f = g \cdot h$, with g, h non-constant polynomials in $\mathbb{Z}_p[X]$ (resp. $\mathbb{Q}_p[X]$).

0.3. *If $f \in \mathbb{Z}_p[X]$, then f is irreducible in $\mathbb{Z}_p[X]$ if and only if it is irreducible in $\mathbb{Q}_p[X]$.*

The non-constant polynomials $f, g \in \mathbb{Z}_p[X]$ are said to be *relatively prime* whenever, if $h \in \mathbb{Z}_p[X]$ and h divides f and g, then $\deg(h) = 0$.

0.4. *If $f \in \mathbb{Z}_p[X]$ is non-constant and primitive, and if f does not divide the non-constant polynomial $g \in \mathbb{Z}_p[X]$, then f and g are relatively prime.*

0.5. *If $f, g \in \mathbb{Z}_p[X]$ are non-constant and relatively prime, then there exist polynomials $s, t \in \mathbb{Z}_p[X]$ such that $s \cdot f + t \cdot g$ is a non-zero element of $\mathbb{Z}_p$.*

0.6. *If $f, g, h \in \mathbb{Z}_p[X]$, if f is irreducible and if f divides $g \cdot h$, then either f divides g or f divides h.*

0.7. If $g, h \in \mathbb{Z}_p[X]$ are non-constant and relatively prime, if g or h is primitive and both g and h divide f, then $g \cdot h$ *divides* f.

0.8. Every non-zero polynomial $f \in \mathbb{Z}_p[X]$ may be written as a product $f = ag_1 \cdots g_n$, where $a \in \mathbb{Z}_p$, and $g_1, \ldots, g_n \in \mathbb{Z}_p[X]$ are primitive irreducible polynomials with $n \geqslant 0$. Moreover, a and $g_1, \ldots, g_n$ are uniquely defined up to a unit in $\mathbb{Z}_p$.

1 The resultant and the discriminant

Let A be any integral domain, and let

$$\begin{aligned} f &= a_0X^m + a_1X^{m-1} + \ldots + a_m, \\ g &= b_0X^n + b_1X^{n-1} + \ldots + b_n, \end{aligned}$$

with $a_i, b_j \in A$, $a_0b_0 \neq 0$, and $m, n > 0$.

Consider the following matrix (*the eliminant of* f, g), of $m+n$ rows and columns:

$$E\ell(f,g) = \begin{pmatrix} a_0 & a_1 & \cdots & a_m & 0 & 0 & \cdots & 0 & 0 & \cdots \\ 0 & a_0 & \cdots & a_{m-1} & a_m & 0 & \cdots & 0 & 0 & \cdots \\ & & \ddots & & & & & & & \\ b_0 & b_1 & \cdots & \cdots & \cdots & \cdots & \cdots & b_n & 0 & \cdots \\ 0 & b_0 & \cdots & \cdots & \cdots & \cdots & \cdots & b_{n-1} & b_n & \cdots \\ & & \ddots & & & & & & & \end{pmatrix}$$

with n rows with the coefficients a_i and m rows with the coefficients b_j.

The determinant of this matrix is called the *resultant* of f, g and denoted by $R(f,g)$; it is an element of A.

Clearly, $R(f,f) = 0$. We extend this definition, letting $R(f, b_0) = b_0^m$ (where $b_0 \in A$, $m = \deg(f) > 0$)) and $R(a_0, g) = a_0^n$ (where $a_0 \in A$, $n = \deg(g) > 0$).

If A' is a domain containing A, and if $f, g \in A[X]$ then the resultant $R(f,g)$ is the same, whether f, g are considered in $A[X]$ or in $A'[X]$.

Let $f' = \sum_{i=0}^{m-1}(m-1)a_iX^{m-1-i}$, that is f' is the derivative of the polynomial f.

If $f, g \in A[X]$ are non-constant polynomials and g^k divides f, then g^{k-1} divides f'. In particular, if $a \in A$ is a root of order $k \geqslant 1$ of f, then a is a root of order at least $k-1$ of f'.

Thus, if $f = a_0X + a_1$ and $a_0 \neq 0$, then $\operatorname{discr}(f) = a_0$. A multiple of a_0, the *discriminant of* f, denoted by $\operatorname{discr}(f)$, is defined by the relation

$$R(f, f') = a_0 \operatorname{discr}(f).$$

The following properties of the resultant and discriminant are well known and their proofs may be found in [1], [2].

Proposition 1.1. *If* $f, g, h, k \in A[X]$, *then:*

(i) $R(g,f) = (-1)^{mn}R(f,g)$.

(ii) If $\deg(f) \leqslant \deg(g)$ *then* $R(f,g) = R(f, g+fh)$ *where* $\deg(f) + \deg(h) \leqslant \deg(g)$.

(iii) The equations

$$\begin{aligned} R(hk,g) &= R(h,g)\cdot R(k,g),\\ R(g,hk) &= R(g,h)\cdot R(g,k), \end{aligned}$$

are satisfied.

(iv) $R(f^s,g) = [R(f,g)]^s$ *for every integer* $s \geqslant 1$.

(v) $R((X-a)^s,g) = [g(a)]^s$ *where* $a \in A$, $s \geqslant 1$.

(vi) If $f = a_0\prod_{i=1}^m (X-\alpha_i)$ *and* $g = b_0\prod_{j=1}^n (X-\beta_j)$ *then*

$$\begin{aligned} R(f,g) &= a_0^n b_0^m \prod_{i=1}^{m}\prod_{j=1}^{n}(\alpha_i-\beta_j)\\ &= a_0^n \prod_{i=1}^{m} g(\alpha_i)\\ &= (-1)^{mn} b_0^m \prod_{j=1}^{n} f(\beta_j). \end{aligned}$$

(vii) If $f = a_0\prod_{i=1}^m (X-\alpha_i)$, *then*

$$\operatorname{discr}(f) = (-1)^{m(m-1)/2} a_0^{2m-1} \prod_{i<j}(\alpha_i-\alpha_j)^2.$$

(viii) If $f = hk$, $\deg(h) = r$, *and* $\deg(k) = s$, *then*

$$\operatorname{discr}(f) = (-1)^{rs}\operatorname{discr}(h)\operatorname{discr}(k)[R(h,k)]^2.$$

Proposition 1.2.

(i) If $f, g \in A[X]$ *are non-constant and* $R(f,g) \neq 0$, *then* f, g *are relatively prime.*

(ii) If $A = K$ *is a field, and if* $f, g \in K[X]$ *are relatively prime, then* $R(f,g) \neq 0$.

Proof

(i) Assume that f, g have a common non-constant factor $h \in A[X]$. So $f = hf_1$ and $g = hg_1$. By (1.1)(iii),

$$R(f,g) = R(h,h)\cdot R(h,g_1)R(f_1,h)R(f_1,g_1) = 0.$$

(ii) Assume that $f, g \in K[X]$ are relatively prime. By Bézout's theorem, there exists $f_1, g_1 \in K[X]$ such that $g_1 f + f_1 g = 1$; in particular, $\deg(g_1 f) = \deg(f_1 g)$. By (1.1)(iii),

$$R(g_1 f, f_1 g) = R(g_1, f_1) \cdot R(g_1, g) \cdot R(f, f_1) \cdot R(f, g).$$

If $R(f, g) = 0$, then $R(g_1 f, f_1 g) = 0$. However, by (1.1)(ii),

$$R(g_1 f, f_1 g) = R(g_1 f, 1 - g_1 f) = R(g_1 f, 1) = 1,$$

which is a contradiction. ◊ ◊

Now we shall consider the resultant and the discriminant of polynomials in $\mathbb{Z}_p[X]$.

Proposition 1.3. *Let f, g be non-constant polynomials in $\mathbb{Z}_p[X]$. Then the following conditions are equivalent:*

(a) There exists a non-constant polynomial $h \in \mathbb{Z}_p[X]$ which divides both f and g.

(b) There exist non-zero polynomials $f_1, g_1 \in \mathbb{Z}_p[X]$ such that $\deg(f_1) < \deg(f)$, $\deg(g_1) < \deg(g)$ *and*

$$g_1 \cdot f + f_1 \cdot g = 0.$$

Proof
(a) $\Longrightarrow$ (b): We assume that $f = h \cdot f_1$, and $g = -h \cdot g_1$, with $f_1, g_1 \in \mathbb{Z}_p[X]$; then $\deg(f_1) < \deg(f)$, $\deg(g_1) < \deg(g)$ and $g_1.f + f_1.g = 0$.
(b) $\Longrightarrow$ (a): Conversely, we assume that there exist polynomials $f_1, g_1 \in \mathbb{Z}_p[X]$, such that $\deg(f_1) < \deg(f)$, $\deg(g_1) < \deg(g)$ and $g_1 f + f_1 g = 0$. If f, g are relatively prime, then by (0.5) there would exist polynomials $s, t \in \mathbb{Z}_p[X]$ such that $s \cdot f + t \cdot g = c \in \mathbb{Z}_p$, with $c \neq 0$. Eliminating g from the above relations, we obtain $f(sf_1 - tg_1) = cf_1$ where $\deg(f_1) < \deg(f)$, which is impossible. ◊ ◊

Proposition 1.4. *In order that $f = \sum_{i=0}^{m} a_i X^{m-i}$ and $g = \sum_{j=0}^{n} b_j X^{n-j}$ (where $m, n > 0$, and $f, g \in \mathbb{Z}_p[X]$) have a common non-constant factor, it is necessary and sufficient that $R(f, g) = 0$.*

Proof It was seen in (1.2) that if f, g have a common non-constant factor, then $R(f, g) = 0$.

Conversely, by (1.3) it is equivalent to show the existence of non-zero polynomials $f_1, g_1 \in \mathbb{Z}_p[X]$, $f_1 = \sum_{i=0}^{m-1} c_i X^{m-1-i}$, $g_1 = \sum_{i=0}^{n-1} d_i X^{n-1-i}$, such that $g_1.f + f_1.g = 0$ (it is not excluded that $c_0 = d_0 = 0$). This relation is equivalent to the following system of $m+n$ equations in the unknown quantities $c_0, \ldots, c_{m-1}, d_0, \ldots, d_{m-1}$ (obtained by equating to zero the coefficients of the powers of X):

$$\begin{cases} d_0a_0 + c_0b_0 & = 0 \\ d_0a_1 + d_1a_0 + c_0b_1 + c_1b_0 & = 0 \\ d_0a_2 + d_1 + d_2a_0 + c_0b_2 + c_1b_1 + c_2b_0 & = 0 \\ & \vdots \end{cases}$$

This homogeneous linear system has a non-trivial solution in $\mathbb{Q}_p$ if and only if the determinant of its matrix vanishes, or equivalently, the determinant of the matrix obtained after exchanging rows and columns vanishes; in other words, $R(f,g) = 0$. Now we finish the proof by noticing that if there exists a non-trivial solution in $\mathbb{Q}_p$, by multiplying by the common denominator of these elements, we obtain a non-trivial solution in $\mathbb{Z}_p$. ◊ ◊

Proposition 1.5. *Let $f, g \in \mathbb{Z}_p[X]$ be relatively prime non-constant polynomials, such that $\nu_p(R(f,g)) = \rho$. Then every non-zero polynomial $h \in \mathbb{Z}_p[X]$, such that $\nu_p(h) \geqslant \rho$, and $\deg(h) < \deg(f) + \deg(g)$, may be written in a unique way as $h = g_1 \cdot f + f_1 \cdot g$, where $f_1, g_1 \in \mathbb{Z}_p[X]$, $\nu_p(f_1) \geqslant \nu_p(h) - \rho$, $\nu_p(g_1) \geqslant \nu_p(h) - \rho$, $\deg(f_1) < \deg(f)$, and $\deg(g_1) < \deg(g)$.*

Proof Let

$$\begin{aligned} f &= \sum_{i=0}^{m} a_i X^{m-i}, \\ g &= \sum_{i=0}^{n} b_i X^{n-i}, \\ h &= \sum_{i=0}^{m+n-1} e_i X^{m+n-i}. \end{aligned}$$

We want to determine $f_1 = \sum_{i=0}^{m-1} c_i X^{m-1-i}$ and $g_1 = \sum_{i=0}^{n-1} d_i X^{n-1-i}$ in $\mathbb{Z}_p[X]$ such that $h = g_1 \cdot f + f_1 \cdot g$. Comparing the coefficients of X in both sides of the above relation, we obtain a linear system of $m+n$ equations in the $m+n$ unknown quantities c_i, d_j, whose determinant is exactly $R(f,g)$.

Since f, g are relatively prime, by (1.4) we have $R(f,g) \neq 0$; hence the above system has unique solution.

The coefficients c_i, d_j may be computed by Cramer's rule; their numerators are linear forms in the e_i with coefficients in $\mathbb{Z}_p$ (because $f, g \in \mathbb{Z}_p[X]$), and their denominators are equal to $R(f,g)$. From $\nu_p(e_i) \geqslant \nu_p(h) \geqslant 0 = \nu_p(R(f,g))$, it follows that $\nu_p(c_i) \geqslant 0$, $\nu_p(d_j) \geqslant 0$, so that $f_1, g_1 \in \mathbb{Z}_p[X]$ and $\nu_p(f_1) \geqslant \nu_p(h) - \rho$, $\nu_p(g_1) \geqslant \nu_p(h) - \rho$. ◊ ◊

Proposition 1.6. *Let $g \in \mathbb{Z}_p[X]$ be a non-constant polynomial. In order that there exists a non-constant polynomial $g \in \mathbb{Z}_p[X]$ such that g^2 divides f, it is necessary and sufficient that* $\operatorname{discr}(f) = 0$.

Proof In fact, if g^2 divides f, then g divides f and f', hence $\operatorname{discr}(f) = R(f, f') = 0$. Conversely, if $\operatorname{discr}(f) = 0$, by (1.3) there exists a non-constant polynomial $f \in \mathbb{Z}_p[X]$ dividing f and f'; by (0.8), we may assume that g is irreducible. We have $f = g \cdot h$, hence $f' = g' \cdot h + g \cdot h'$; since g divides f', it follows that g divides $g'.h$; from $\deg(g') < \deg(g)$ we see that g does not divide g', hence g divides h (by (0.6)), and so g^2 divides f. ◊ ◊

We shall now investigate the behaviour of the resultant $R(f, g)$ when f, g are replaced by sufficiently close polynomials, relative to the metric defined by the valuation ν_p on $\mathbb{Q}_p(X)$.

Proposition 1.7. *If $f, g, f_1, g_1 \in \mathbb{Z}_p[X]$ are non-constant polynomials and $\nu_p(f_1 - f) \geqslant \alpha$, $\nu_p(g_1 - g) \geqslant \beta$, then $\nu_p(R(f_1, g_1) - R(f, g)) \geqslant \min\{\alpha, \beta\}$.*

Proof Let $a \in \mathbb{Z}_p$ be such that $\nu_p(a) = \min\{\nu_p(f_1 - f), \nu_p(g_1 - g)\}$; then $f_1 = f + ah$ and $g_1 = g + ak$, where $h, g \in \mathbb{Z}_p[X]$. Thus $R(f_1, g_1) = R(f + ah, g + ak)$. Writing the eliminating matrix between $f + ah$, $g + ak$, and computing the determinant, we obtain $R(f, g) + as$, where $s \in \mathbb{Z}_p$ is a certain sum of products of elements equal to a or to coefficients of f, h, g, k. Thus

$$\nu_p(R(f_1, g_1) - R(f, g)) \geqslant \nu_p(a) \geqslant \min\{\alpha, \beta\}.$$

◊ ◊

Proposition 1.8. *With the above notations, if $f, f_1 \in \mathbb{Z}_p[X]$ are non-constant polynomials and $\nu_p(f - f_1) \geqslant \alpha$, then $\nu_p(\operatorname{discr}(f) - \operatorname{discr}(f_1)) \geqslant \alpha$.*

Proof In fact, since $\operatorname{discr}(f) = R(f, f')$ and $\operatorname{discr}(f_1) = R(f_1, f_1')$, we have only to remark that $\nu_p(f_1 - f) \geqslant \alpha$ implies $\nu_p(f_1' - f') \geqslant \alpha$. Indeed, if $m = \max\{\deg(f), \deg(f_1)\}$ and $f = \sum_{i=0}^{m} a_i X^{m-i}, f_1 = \sum_{i=0}^{m} b_i X^{m-i}$,, then $f' = \sum_{i=0}^{m-1} a_i X^{m-1-i}$ and $f_1' = \sum_{i=0}^{m-1} (m-1) b_i X^{m-1-i}$; thus

$$\nu_p((m-1)(b_i - a_i)) \geqslant \nu_p(b_i - a_i)$$

for every $i = 0, \ldots, m-1$, and so

$$\nu_p(f_1' - f') \geqslant \min\{\nu_p(b_i - a_i) : i = 0, ..., m-1\} \geqslant \nu_p(f_1 - f) \geqslant \alpha.$$

◊ ◊

We have also at once:

Proposition 1.9. *If $f, g \in \mathbb{Z}_p[X]$ and $\bar{f} = f \bmod p$, $\bar{g} = g \bmod p$, then the resultant of $\bar{f}$ and $\bar{g}$ (computed in $\mathbb{F}_p[X]$)) is $R(\bar{f}, \bar{g}) = \overline{R(f, g)}$ and the discriminant of $\bar{f}$ (computed in $\mathbb{F}_p[X]$)) is $\operatorname{discr}(\bar{f}) = \overline{\operatorname{discr}(f)}$.*

We say that the monic non-constant polynomials $f, g \in \mathbb{Z}_p[X]$ are *relatively prime modulo p* when $\bar{f}, \bar{g}$ are relatively prime polynomials in $\mathbb{F}_p[X]$. Similarly, f is said to be *irreducible modulo p* whenever $\bar{f}$ is an irreducible polynomial of $\mathbb{F}_p[X]$. Every polynomial $f \in \mathbb{Z}_p[X]$ is congruent modulo p to a product of polynomials in $\mathbb{Z}_p[X]$, which are irreducible modulo p, and are uniquely defined modulo p.

With these definitions, we have:

Proposition 1.10. *$f, g \in \mathbb{Z}_p[X]$ are relatively prime modulo p if and only if the resultant is a unit in $\mathbb{Z}_p$, i.e. if and only if $\nu_p(R(f, g)) = 0$.*

Proof By definition, f, g are relatively prime modulo p when $\bar{f}, \bar{g}$ are relatively prime polynomials in $\mathbb{F}_p[X]$; by (1.2), this means that $R(\bar{f}, \bar{g}) \neq 0$; by (1.9), this is equivalent to $\overline{R(f, g)} \neq 0$, that is p does not divide $R(f, g)$, or equivalently, $\nu_p(R(f, g)) = 0$. ◊ ◊

Proposition 1.11. *Let $f, g \in \mathbb{Z}_p[X]$ be irreducible polynomials modulo p. Then p divides $R(f, g)$ if and only if $f \equiv g \bmod p$.*

Proof In fact, p divides $R(f, g)$ exactly when f, g are not relatively prime modulo p; hence there exists a non-constant polynomial $h \in \mathbb{Z}_p[X]$ such that $f \equiv h \cdot f_1 \bmod p$, $g \equiv h \cdot g_1 \bmod p$; by hypothesis, we must have $f \equiv h \bmod p$, $g \equiv h \bmod p$, hence $f \equiv g \bmod p$. The converse is trivial. ◊ ◊

We observe that if $f = \sum_{i=0}^{m} \in \mathbb{Z}_p[X]$ and f' is a multiple of p, then $f \equiv \sum_{i \geqslant 0} a_{pi} X^{pi} \bmod p$. Indeed, the coefficients of f' are $j a_j$; thus if p divides f', then p divides a_j when p does not divide j. Therefore

$$f \equiv \left(\sum_{i \geqslant 0} a_{pi} X^{pi} \right)^p \bmod p.$$

In particular, if f is irreducible modulo p, then $\bar{f}' \neq 0$ and hence we may consider the discriminant of f modulo p.

Proposition 1.12. *If $f \in \mathbb{Z}_p[X]$ is irreducible modulo p, then p does not divide* $\operatorname{discr}(f)$.

Proof We write $f = f_1 + pf_2$ where all the coefficients of f_1 are not multiples of p. Then $f' = f_1' + pf_2'$ and by (1.8), $\mathrm{discr}(f) \equiv \mathrm{discr}(f_1) \bmod p$. If p divides $\mathrm{discr}(f)$, then p divides $\mathrm{discr}(f_1) = R(f_1, f_1')$. By (1.10), there exists $h \in \mathbb{Z}_p[X]$ such that $\bar{h}$ is non-constant and $\bar{h}$ is a common factor of $\bar{f}_1$, $\bar{f}_1'$. Thus $\bar{f}_1 = \bar{h} \cdot \bar{g}$, $\bar{f}_1' = \bar{h} \cdot \bar{k}$ with $g, k \in \mathbb{Z}_p[X]$. Since $\bar{f}_1 = \bar{f}$ is irreducible, then $\bar{g} = \bar{c}$ with $c \in \mathbb{Z}_p$. So $\bar{c} \cdot \bar{f}_1' = \bar{f}_1 \cdot \bar{k}$. Therefore

$$\deg(\bar{f}_1') \leqslant \deg(f_1') = \deg(\bar{f}_1) \leqslant \deg(\bar{f}_1 \cdot \bar{k}) = \deg(\bar{f}_1'),$$

which is absurd. ◊ ◊

Proposition 1.13. *Let $f \in \mathbb{Z}_p[X]$ be such that $\bar{f}$ is not constant. Then f has a multiple irreducible factor modulo p if and only if p divides* $\mathrm{discr}(f)$.

Proof We have $f \equiv g_1 g_2 \cdots g_n \bmod p$, where $g_1, g_2, \ldots, g_n$ are irreducible modulo p. Hence, by (1.1)(iii), (1.7) and (1.1)(viii),

$$\begin{aligned} \mathrm{discr}(f) &\equiv \mathrm{discr}(g_1 g_2 \cdots g_n) \\ &= \pm \prod_{i=1}^{n} \mathrm{discr}(g_i) \cdot \prod_{i<j} [R(g_i, g_j)]^2 \bmod p. \end{aligned}$$

By (1.13), p does not divide $\mathrm{discr}(g_i)$ for $1 \leqslant i \leqslant n$. Then p divides $\mathrm{discr}(f)$ if and only if there exist indices $i < j$ such that p divides $R(g_i, g_j)$. By (1.11), this means that $g_i \equiv g_j \bmod p$, concluding the proof. ◊ ◊

2 Hensel's Lemma

This very important result, proved by Hensel [3] in 1908, is certainly the *raison d'être* of the p-adic numbers. It asserts the existence, under appropriate conditions, of p-adic roots of polynomials. We shall prove here Hensel's Lemma in its strong form:

Theorem 2.1. *Let $F, g, h \in \mathbb{Z}_p[X]$ be such that:*

(i) $\deg(g) = m > 0$, $\deg(h) = n > 0$, $\deg(F) = m + n$, *g is monic and* $\deg(F - gh) < \deg(F)$.

(ii) $\nu_p(R(g, h)) = \rho \geqslant 0$.

(iii) $\nu_p(F - gh) = \alpha > 2\rho$.

Then there exist $G, H \in \mathbb{Z}_p[X]$ such that $\nu_p(G - g) \geqslant \alpha - \rho$, $\nu_p(H - h) \geqslant \alpha - \rho$, $\deg(G) = \deg(g)$, $\deg(H) = \deg(h)$, G is monic H, h have the same leading coefficient, and finally, $F = G \cdot H$.

Proof We shall prove the following assertion, for $j \geqslant 0$:

($*$) If $g, h \in \mathbb{Z}_p[X]$, $\deg(g) = m$, $\deg(h) = n$, g is monic, $\deg(F - gh) < \deg(F)$, $\nu_p(F - gh) \geqslant \alpha + j$, and $\nu_p(R(g, h)) = \rho$, then there exist polynomials $g^*, h^* \in \mathbb{Z}_p[X]$, such that $\deg(g^*) < m$, $\deg(h^*) < n$, $\nu_p(g^*) \geqslant \alpha + j - \rho$, $\nu_p(h^*) \geqslant \alpha + j - \rho$, and $\nu_p(F - (g + g^*)(h - h^*)) \geqslant \alpha + j - 1$.

Indeed, since $\nu_p(R(g, h)) = \rho$ then $R(g, h) \neq 0$. By (1.4), g, h are relatively prime.

We note that $\nu_p(F - gh) \geqslant \alpha + j\alpha\rho$ and $\deg(F - gh) < \deg(F) = \deg(g) + \deg(h)$, and it follows from (1.5) that there exist uniquely defined polynomials $g^*, h^* \in \mathbb{Z}_p[X]$, such that the following hold:

$$F - gh = h^* g + g^* h,$$

$$\nu_p(g^*) \geqslant \nu_p(F - gh) - \rho \geqslant \alpha + j - \rho,$$

$$\nu_p(h^*) \geqslant \nu_p(F - gh) - \rho \geqslant \alpha + j - \rho,$$

$$\begin{aligned} \deg(g^*) &< \deg(g), \\ \deg(h^*) &< \deg(h). \end{aligned}$$

Therefore

$$\begin{aligned} \nu_p(F - (g + g^*)(h + h^*)) &= \nu_p((F - gh) - (h^* g + g^* h) - g^* h^*) \\ &= \nu_p(g^* h^*) \\ &\geqslant 2(\alpha + j - \rho) \end{aligned}$$

$$\begin{aligned} &= (\alpha - 2\rho) + (\alpha + 2j) \\ &\geqslant \alpha + j + 1. \end{aligned}$$

We apply this result, beginning with $g = g_0$, $h = h_0$, $j \geqslant 0$ and letting $g_1 = g_0 + g_0^*$, $h_1 = h_0 + h_0^*$. Then, we apply the result for g_1, h_1, $j = 1$, obtaining $g_2 = g_1 + g_1^*$, $h_2 = h_1 + h_1^*$ and so on.

We have to note that

$$\nu_p(R(g_{j+1}, h_{j+1})) = \nu_p(R(g_j, h_j)) = \rho$$

for every $j \geqslant 0$, because

$$\begin{aligned} \nu_p[R(g_{j+1}, h_{j+1}) - R(g_j, h_j)] &\geqslant \min\{\nu_p(g^*), \nu_p(h^*)\} \\ &\geqslant \alpha + j - \rho > \rho \end{aligned}$$

as follows from (1.7).

Thus, we have the sequences of polynomials $(g_j)_{j\geqslant 0}$ and $(h_j)_{j\geqslant 0}$ such that $\deg(g_j) = m$, $\deg(h_j) = n$, each g_j is monic, h_j and h have the same leading coefficient, and finally

$$\begin{aligned} \nu_p(g_{j+1} - g_j) &\geqslant \alpha + j - \rho, \\ \nu_p(h_{j+1} - h_j) &\geqslant \alpha + j - \rho. \end{aligned}$$

Thus $(g_i)_{j\geqslant 0}$ and $(h_j)_{j\geqslant 0}$ are Cauchy sequences of polynomials of degree m, n respectively. This means that if

$$\begin{aligned} g_j &= \sum_{i=0}^{m} b_{ij} X^i, \\ h_j &= \sum_{i=0}^{n} c_{ij} X^i, \end{aligned}$$

then the sequences $(b_{ij})_{j\geqslant 0}$, $(c_{ij})_{j\geqslant 0}$ (for every i), are Cauchy sequences in $\mathbb{Q}_p$. Since $\mathbb{Q}_p$ is complete, let $b_i = \lim b_{ij}$, $c_i = \lim c_{ij}$ and $G = \sum_{i=0}^{m} b_i X^i$, $H = \sum_{i=0}^{n} c_i X^i$.

Then $\nu_p(G - g_j) \geqslant \alpha + j - \rho$, because

$$G - g_j = \lim_{s \longrightarrow \infty} \left(\sum_{i=0}^{s} g_{j+1}^* \right)$$

and

$$\nu_p \left(\sum_{i=0}^{s} g_{j+i}^* \right) \geqslant \alpha + j - \rho$$

for every $s \geqslant 1$. Similarly, $\nu_p(H - h_j) \geqslant \alpha + j - \rho$ for every $j \geqslant 0$.

Finally,

$$\begin{aligned}\nu_p(F - GH) &= \nu_p[(F - g_j h_j) + (g_j - G)H + g_j(h_j - H)] \\ &\geqslant \min\{\nu_p(F - g_j h_j), \nu_p(g_j - G) + \nu_p(H), \nu_p(g_j) + \nu_p(h_j - H)\} \\ &\geqslant \alpha + j - \rho\end{aligned}$$

for every $j \geqslant 0$. It follows that $F = GH$ with G monic, and H, h having the same leading coefficient. ◊ ◊

Now we give Hensel's Lemma in its more customary form:

Theorem 2.2. *Let $F, g, h \in \mathbb{Z}_p[X]$ be such that:*

(i) $\deg(g) = m > 0$, $\deg(h) = n > 0$, $\deg(F) = m + n$, *g is monic and* $\deg(F - gh) < \deg(F)$.

(ii) g, h are relatively prime modulo p.

(iii) $F \equiv g \cdot h \bmod p$.

Then there exist polynomials $G, H \in \mathbb{Z}_p[X]$ such that $G \equiv g \bmod p$, $H \equiv h \bmod p$, $\deg(G) = \deg(g)$, $\deg(H) = \deg(h)$, G is monic, H, h have the same leading coefficient and $F = G \cdot H$.

Proof This is an immediate corollary of the preceding result. Indeed, by (1.10), $\nu_p(R(g, h)) = 0$. Since $\nu_p(F - gh) \geqslant 1$, the above result may be applied. ◊ ◊

Another commonly encountered form of Hensel's lemma concerns the lifting of roots modulo p.

Theorem 2.3. *Let $F \in \mathbb{Z}_p[X]$ with* $\deg(F) \geqslant 1$, *let $a \in \mathbb{Z}_p$ be a simple root of the congruence $F(X) \equiv 0 \bmod p$. Then there exists $b \in \mathbb{Z}_p$ such that $\bar{b} = \bar{a}$ and $F(b) = 0$.*

Proof By hypothesis $F \equiv (X - a)h \bmod p$ where $h(a) \not\equiv 0 \bmod p$. So $X - a, h$ are relatively prime modulo p. By (2.2) $F = GH$, with G monic, $\deg(G) = 1$, $\bar{G} = X - a$, thus $G = X - b$ with $b \in \mathbb{Z}_p$, $\bar{b} = \bar{a}$, and therefore $F(b) = 0$. ◊ ◊

3 The local Fermat problem

In this section we shall consider Fermat's equation

$$X^p + Y^p = Z^p \quad \text{(for } p \geqslant 2 \text{ and } p \text{ a prime)} \tag{3.1}$$

and investigate solutions in the fields $\mathbb{Q}_q$ of q-adic numbers. Our main result is

Theorem 3.1. *For every prime q and every prime p, Fermat's equation (3.1) has a non-trivial solution in the field $\mathbb{Q}_q$.*

Proof If $p = 2$, the equation $X^2+Y^2 = Z^2$ has, as is well known, infinitely many non-trivial solutions in integers (they are the so-called *Pythagorean triples*); for example, the triples of integers

$$(3,4,5),\ (5,12,13),\ \ldots.$$

These are, of course, solutions in $\mathbb{Q}_p$ for every prime q.

Now we assume that $p \neq 2$; it is equivalent to consider the equation

$$X^p + Y^p + Z^p = 0 \quad \text{(for } p > 2 \text{ and } p \text{ a prime.)} \tag{3.2}$$

First case: $q \neq p$.

Let $F(X) = X^p + q^p - 1$. Then

$$X^p + q^p - 1 \equiv (X-1)(X^{p-1} + X^{p-2} + \cdots + X + 1) \bmod q.$$

Since 1 mod q is not a root of $X^{p-1}+X^{p-2}+\cdots+X+1 \equiv 0 \bmod q$, then by Hensel's lemma (2.3) there exists a q-adic integer α, such that $\alpha \equiv 1 \bmod q$ and $\alpha^p + q^p + (-1)^p = 0$.

Second case: $q = p$.

The proof is similar with an appeal to the stronger form of Hensel's Lemma. Again, let

$$\begin{aligned} F(X) &= X^p + p^p - 1, \\ G_0(X) &= X - 1, \\ H_0(X) &= X^{p-1} + X^{p-2} + \cdots + X + 1. \end{aligned}$$

Then $F(X) \equiv G_0(X)H_0(X) \bmod p^p$.

The resultant

$$R = R(G_0(X), H_0(X)) = (-1)^{p-1}H_0(1) = p,$$

so its p-adic value is $\nu_p(R) = 1$. Noting that $p \geqslant 3 > 2\nu_p(R)$, we may apply Hensel's Lemma (in the form indicated in (2.1)). Thus, there exist monic polynomials $G(X)$, $H(X) \in \mathbb{Z}_p[X]$ such that

$$G(X) \equiv G_0(X) \bmod p^{p-1}$$

$$H(X) \equiv H_0(X) \bmod p^{p-1}$$

and

$$F(X) = G(X)H(X).$$

Thus $G(X) = X - \alpha$ where we have $\alpha \in \mathbb{Z}_p$, $\alpha \equiv 1 \bmod p^{p-1}\mathbb{Z}_p$ and $\alpha^p + p^p + (-1)^p = 0$. ◊ ◊

Obviously for every $p \geqslant 2$ the equation 3.1 has a non-trivial solution in real numbers. On the other hand, according to Fermat's conjecture (already proved for many values of p) the equation 3.1 has only trivial solutions in the rational numbers. We conclude that the so-called 'local to global principle' does not hold generally for the equation 3.1, for example when $p = 3$.

A related and interesting question is the determination of solutions (α, β, γ) of (3.1) where α, β, γ are q-adic units (for $q \geqslant 2$, q a prime).

Theorem 3.2. *Let q, p be distinct primes, $p \neq 2$, $q \geqslant 2$. Then the following conditions are equivalent:*

a) There exist units $\alpha, \beta, \gamma \in \mathbb{Z}_p$ such that

$$\alpha^p + \beta^p + \gamma^p = 0.$$

b) There exist integers x_0, y_0, z_0, not multiples of q, such that

$$x_0^p + y_0^p + z_0^p \equiv 0 \bmod q.$$

Proof

(a) $\Longrightarrow$ (b): We write

$$\begin{cases} \alpha &= x_0 + \alpha' q \\ \beta &= y_0 + \beta' q \\ \gamma &= z_0 + \gamma' q \end{cases}$$

with $x_0, y_0, z_0 \in \mathbb{Z}$, $\alpha', \beta', \gamma' \in \mathbb{Z}_p$. Since α, β, γ are units in $\mathbb{Z}_p$ then $q \nmid x_0 y_0 z_0$. From $\alpha^p + \beta^p + \gamma^p = 0$ it follows that

$$\begin{aligned} 0 = \alpha^p + \beta^p + \gamma^p &= (x_0 + \alpha' q)^p + (y_0 + \beta' q)^p + (z_0 + \gamma' q)^p \\ &\equiv x_0^p + y_0^p + z_0^p \bmod q. \end{aligned}$$

(b) $\Longrightarrow$ (a): First we shall prove the existence of sequences of integers $(x_n)_{n \geqslant 0}$, $(y_n)_{n \geqslant 0}$, $(z_n)_{n \geqslant 0}$ with $q \nmid x_n y_n z_n$ (for every n), such that

$$\begin{cases} x_{n+1} &\equiv x_n \bmod q^{n+1} \\ y_{n+1} &\equiv y_n \bmod q^{n+1} \\ z_{n+1} &\equiv z_n \bmod q^{n+1} \end{cases}$$

and

$$x_n^p + y_n^p + z_n^p \equiv 0 \bmod q^{n+1}.$$

We begin with the given integers x_0, y_0, z_0 and proceed by induction, assuming that $x_i, y_i, z_i \in \mathbb{Z}$ for $0 \leqslant i \leqslant m$) have already been found, satisfying the required conditions.

Thus $x_m^p + y_m^p + z_m^p \equiv r'q^{m+1} \bmod q$, where $r' \in \mathbb{Z}$. Since $q \nmid z_m$ there exists an integer r satisfying the congruence

$$rpz_m^{p-1} \equiv r' \bmod q.$$

Let $x_{m+1} = x_m$, $y_{m+1} = y_m$ and $z_{m+1} = z_m - rq^{m+1}$. Then

$$z_{m+1}^p = (z_m - rq^{m+1})^p \equiv z_m^p - prq^{m+1}z_m^{p-1} \bmod q^{m+2}$$

and

$$\begin{aligned} x_{m+1}^p + y_{m+1}^p + z_{m+1}^p &\equiv x_m^p + y_m^p + z_m^p - prq^{m+1}z_m^{p-1} \\ &\equiv (r' - prz_m^{p-1})q^{m+1} \equiv 0 \bmod q^{m+2}. \end{aligned}$$

The sequences $(x_n)_{n\geqslant 0}$, $(y_n)_{n\geqslant 0}$, $(z_n)_{n\geqslant 0}$ are convergent in the field $\mathbb{Q}_q$ of q-adic numbers. Let $\alpha = \lim x_n$, $\beta = \lim y_n$, $\gamma = \lim z_n$. Then $\alpha, \beta, \gamma \in \mathbb{Z}_q$. Also, from the equation

$$x_n^p + y_n^p + z_n^p \equiv 0 \bmod q^{n+1}$$

it follows that $\alpha^p + \beta^p + \gamma^p = 0$. Finally, since $q \nmid x_n y_n z_n$ for every $n \geqslant 0$, then α, β, γ are q-adic units. ◊ ◊

From the above proof, we note that $\alpha = x_0$, $\beta = y_0$.

We give the corresponding result when $q = p$:

Theorem 3.3. *Let p be an odd prime. The following conditions are equivalent:*

a) There exist units $\alpha, \beta, \gamma \in \mathbb{Z}_p$ such that

$$\alpha^p + \beta^p + \gamma^p = 0.$$

b) There exist integers x_0, y_0, z_0, not multiples of p, such that

$$x_0^p + y_0^p + z_0^p \equiv 0 \bmod p^2.$$

Proof

(a) $\Longrightarrow$ (b): The proof is like that of (3.2). Let

$$\begin{cases} \alpha &= x_0 + \alpha' p \\ \beta &= y_0 + \beta' p \\ \gamma &= z_0 + \gamma' p \end{cases}$$

where $x_0, y_0, z_0 \in \mathbb{Z}$, $p \nmid x_0 y_0 z_0$ and $\alpha', \beta', \gamma' \in \mathbb{Z}_p$. Then

$$\begin{aligned} 0 = \alpha^p + \beta^p + \gamma^p &= (x_0 + \alpha' p)^p + (y_0 + \beta' p)^p + (z_0 + \gamma' p)^p \\ &\equiv x_0^p + y_0^p + \gamma_0^p \bmod p^2. \end{aligned}$$

(b) $\Longrightarrow$ (a): Again, the proof is similar to that of (3.2). Assume that $x_i, y_i, z_i \in \mathbb{Z}$ with $p \nmid x_i y_i z_i$ for $0 \leqslant i \leqslant m$) have already been found, satisfying

$$\begin{aligned} x_i^p + y_i^p + z_i^p &\equiv 0 \bmod p^{i+2} \\ x_i &\equiv x_{i-1} \bmod p^i \\ y_i &\equiv y_{i-1} \bmod p^i \\ z_i &\equiv z_{i-1} \bmod p^i \end{aligned}$$

for $1 \leqslant i \leqslant m$.

We write $x_m^p + y_m^p + z_m^p = r' p^{m+2}$, with $r \in \mathbb{Z}$. Since $p \nmid z_m$, there exists $r \in \mathbb{Z}$ such that $r z_m^{p-1} \equiv r' \bmod p$.

Let $x_{m+1} = x_m$, $y_{m+1} = y_m$ and $z_{m+1} = z_m - r p^{m+1}$. Then

$$\begin{aligned} z_{m+1}^p &= z_m^p - r p^{m+2} z_m^{p-1} + \binom{p^{m+1}}{2} r^2 p^{2(m+1)} z_m^{p-2} - \cdots \\ &\equiv z_m^p - r p^{m+2} z_m^{p-1} \bmod p^{m+3} \end{aligned}$$

since p^{m+3} divides

$$\binom{p^{m+1}}{2} p^{2(m+1)}, \binom{p^{m+1}}{3} p^{3(m+1)}, \ldots .$$

It follows that

$$\begin{aligned} x_{m+1}^p + y_{m+1}^p + z_{m+1}^p &\equiv x_m^p + y_m^p + (z_m^p - r p^{m+2} z_m^{p-1}) \\ &\equiv (r' - r z_m^{p-1}) p^{m+2} \equiv 0 \bmod p^{m+3}. \end{aligned}$$

Again, the sequences $(x_n)_{n \geqslant p}$, $(y_n)_{n \geqslant 0}$, and $(z_n)_{n \geqslant 0}$ are convergent in $\mathbb{Z}_p$; let $\alpha = \lim x_n$, $\beta = \lim y_n$, $\gamma = \lim z_n$, so α, β, γ are units of $\mathbb{Q}_p$, because $p \nmid x_n y_n z_n$ for every $n \geqslant 0$. Finally, from $x_n^p + y_n^p + z_n^p \equiv 0 \bmod p^{n+2}$, it follows that $\alpha^p + \beta^p + \gamma^p = 0$. ◊ ◊

From the above proof, we note also that $\alpha = x_0$, $\beta = y_0$.

From (3.2) we are led to study the congruence

$$X^p + Y^p + Z^p \equiv 0 \bmod q \tag{3.3}$$

where p, q are distinct primes, and $p \neq 2$.

Let $N(p,q)$ denote the number of triples of integers x, y, z, such that $1 \leqslant x, y, z \leqslant q-1$ and $x^p + y^p + z^p \equiv 0 \bmod q$.

For example, if $p = 3$, Legendre had shown already in 1830 that $N(3,7) = N(3,13) = 0$. In 1880, Pepin determined a formula for $N(3,q)$ when $q \equiv 1 \bmod 3$ and deduced that $N(3,q) > 0$ for $q \geqslant 19$.

It is very easy to see that

If $q \equiv 1 \bmod p$, then $N(p,q) > 0$.

Indeed, let a, b be such that $ap + b(q-1) = 1$. Let x_0, y_0, z_0 be integers, not multiples of q, such that $x_0 + y_0 + z_0 \equiv 0 \bmod q$. Then from $m^{q-1} \equiv 1 \bmod q$, when q does not divide m (Fermat's Little Theorem) it follows that

$$x_0^{ap} + y_0^{ap} + z_0^{ap} \equiv 0 \bmod q,$$

showing the assertion.

In 1909, Dickson found the following upper and lower bounds for $N(p,q)$, where $q \equiv 1 \bmod p$:

$$\begin{aligned}(q-1)\,|q+1-3p-(p-1)(p-2)\sqrt{q}| &< N(p,q)\\ &< (q-1)\,|q+1+3p+(p-1)(p-2)\sqrt{q}|\,.\end{aligned}$$

It follows that if $q \geqslant (p-1)^2(p-2)^2 + 6p - 2$ then $N(p,q) > 0$.

The congruence

$$X^p + Y^p + Z^p \equiv 0 \bmod p^2, \tag{3.4}$$

where $p \neq 2$, has also been extensively studied and this is discussed in [4].

References

1. M. Bôcher, 'Introduction to Higher Algebra', Macmillan, New York (1907) (reprinted in 1947).

2. P.M. Cohn, 'Algebra', Vol. I, J. Wiley, New York (1974).

3. K. Hensel, 'Theorie der algebraischen Zahlen', Tebner, Leipzig, 1908.

4. P. Ribenboim, '13 Lectures on Fermat's Last Theorem', Springer-Verlag, New York (1979).

7.

L-FUNCTIONS AND REPRESENTATION THEORY OF p-ADIC GROUPS

FREYDOON SHAHIDI

Purdue University,
West Lafayette,
IN 47907,
USA.
<shahidi@purdue.math.edu>

Introduction

The modern theory of automorphic forms as envisioned by Langlands [28, 30] gives an equal weight to representation theory of p-adic reductive groups as that of real groups. While representation theory of real groups is near perfection, a similar statement cannot be made about p-adic groups and in fact there is an abundance of unsolved problems of extreme depth, beauty, and importance that must be answered.

The work of Harish-Chandra [11, 45] has already proved that a large part of the theory from real groups carries over, fairly precisely, to the p-adic case. Nothing like this can be said of their discrete series representations and in fact to this date our understanding of them remains poor, except for GL_n and certain other classical groups (see the next paragraph). These are irreducible unitary representations whose matrix coefficients are square integrable modulo the centre of the group. Moreover, they are building blocks for construction of arbitrary irreducible admissible representations of the group. More precisely, using the machinery of parabolic induction every other representation can be constructed from them (cf. [29]).

While the problem of understanding discrete series, especially the supercuspidal ones (Section 5), has been confronted with some success and many interesting results are available (see for example [7, 15, 25, 31, 34, 51]), it is the problem of parabolic induction which will be studied in this article.

In fact, we shall try to explain in a language, as simple as possible, the theory developed in several papers [12, 13, 21, 22, 36, 44, 45] and in particular for the sake of simplicity we restrict ourselves to split groups.

Our approach uses the theory of R-groups (Section 4) and Plancherel measures (Section 3), an approach which seems to be in perfect harmony with the parametrization problem, i.e. the problem of parametrizing representations of the group by means of homomorphism of the Deligne–Weil group into the L-group (Section 2 here, Chapter III of [3], Theorem 3.5 of [36], Theorem 3.4 of [23], and [29]).

According to a conjecture of Langlands (Appendix II of [27]), Plancherel measures are supposed to be ratios of certain local L-functions and root numbers (cf. equation (3.1) of Section 3), and therefore they are also of arithmetic significance (cf. Theorem 3.5 and Corollary 3.6 of [36]). Under an assumption on the inducing representation (Section 5) this conjecture was proved in [36]. Proposition 6.1 of Section 6 and Theorem 8.1 of [36] (Theorem 7.1 of this paper) then explain how this can be used to answer reducibility questions for induced representations, both on and off the unitary axis. The standard normalization of intertwining operators and a sketch of the proof of Theorem 4.1 are also discussed in Section 7.

In Section 8, after discussing a few examples, we consider the symplectic group $\mathrm{Sp}_{2n}(F)$ in $2n$ variables (of rank n) and state a theorem (Theorem 8.1) on the reducibility of representations induced from an irreducible supercuspidal representation of its $\mathrm{GL}_n(F)$ Levi factor. It turns out that the result completely agrees with the parametrization problem discussed above (cf. the introduction of [17]).

We should remark that while in the case of $\mathrm{GL}_{2n}(F)$ with the Levi subgroup $\mathrm{GL}_n(F) \times \mathrm{GL}_n(F)$ the reducibility criterion (in terms of conjectured parametrization) is fairly simple [2, 49], it is much deeper for $\mathrm{Sp}_{2n}(F)$ and its interpretation requires the use of the theory of twisted endoscopy which is now being developed by Kottwitz and Shelstad. In fact, as is evident from the cases of other classical groups, one must expect certain deep results such as character identities stemming from the theory of twisted endoscopy to answer reducibility questions for p-adic groups, a feature which was not confronted at all when similar questions for GL_n were being studied.

Theorem 8.1 is new. Its proof and interpretation in terms of twisted endoscopy will be given in another paper.

Acknowledgements: I would like to thank the editors for their suggestion that an article like this be written for inclusion in this book.

The author was supported by NSF Grants DMS-9000256 and DMS-8610730.

1 Intertwining operators

Let F be a p-adic field of characteristic zero. If O and P are the ring of integers of F and its maximal ideal, we let q be the number of elements in the residue field O/P. We use $|\ |_F$ to denote the absolute value in F for which a prime element ϖ has $|\varpi|_F = q^{-1}$.

Let **G** be a connected reductive algebraic group over F. For simplicity we shall assume that **G** is split, i.e. it has a maximal torus which splits over F. Let **T** be a maximal torus of **G** and choose a Borel subgroup **B** containing **T** whose unipotent radical is denoted by **U**. Then $\mathbf{B} = \mathbf{TU}$. For every group **H** over F, we use H to denote the group $\mathbf{H}(F)$ of F-points of **H**. We refer to [4], [5] and [46] for these definitions.

Let **P** be a parabolic subgroup of **G**, containing **B**. Then its unipotent radical **N** is contained in **U**. Let **M** be a Levi subgroup of **P**. Then $\mathbf{P} = \mathbf{MN}$. Denote by **A** the split component of the centre of **M**, i.e. the largest torus in the centre of **M**. Since **G** is split we may take $K = \mathbf{G}(O)$ as a maximal compact subgroup of G satisfying $G = PK = BK$.

If $X(\mathbf{M}) = \mathrm{Hom}(\mathbf{M}, \mathrm{GL}_1)$ denotes the group of characters of **M**, we let

$$\mathfrak{a} = \mathrm{Hom}(X(\mathbf{M}), \mathbb{R})$$

and $\mathfrak{a}^*_{\mathbb{C}} = \mathfrak{a}^* \otimes_{\mathbb{R}} \mathbb{C}$, where $\mathfrak{a}^* = X(\mathbf{M}) \otimes_{\mathbb{Z}} \mathbb{R} = X(\mathbf{A}) \otimes_{\mathbb{Z}} \mathbb{R}$.

Next, define a homomorphism $H_M: M \to \mathfrak{a}$ by

$$q^{\langle \chi, H_M(m)\rangle} = |\chi(m)|_F$$

for all $\chi \in X(\mathbf{M})$. We extend H_M to a function H_P on P and G by $H_P(mnk) = H_M(m)$.

Let ψ be the set of roots of **T**. Then $\psi = \psi^+ \cup \psi^-$, where ψ^+ is the set of positive roots, i.e. those generating **U**. Let $\Delta \subseteq \psi^+$ be the set of simple roots. We use $\theta \subseteq \Delta$ to denote the subset of roots generating **M**. Let $\rho_{\mathbf{P}}$ be half the sum of roots in **N**. Clearly $\rho_{\mathbf{P}} \in \mathfrak{a}^*$. Finally, let $W(\mathbf{T})$ be the Weyl group of **T** in **G**, i.e. the quotient of the normalizer of **T** in **G** by its centralizer.

Given an irreducible admissible representation $(\sigma, \mathcal{H}(\sigma))$ of M and $\nu \in \mathfrak{a}^*_{\mathbb{C}}$, let

$$I(\nu, \sigma) = \operatorname*{Ind}_{MN\uparrow G}(\sigma \otimes q^{\langle \nu, H_M(\ )\rangle}) \otimes \mathbf{1}$$

be the representation of G induced from σ. The space $V(\nu, \sigma)$ consists of all the smooth (locally constant) functions $f: G \to \mathcal{H}(\sigma)$ satisfying

$$f(mng) = \sigma(m) q^{\langle \nu + \rho_{\mathbf{P}}, H_M(\ )\rangle} f(g),$$

$m \in M$, $n \in N$, and $g \in G$. The group G acts by right translations on $V(\nu, \sigma)$. We set $I(\sigma) = I(0, \sigma)$ and $V(\sigma) = V(0, \sigma)$.

Fix a $\tilde{w} \in W(\mathbf{T})$ such that $\tilde{w}(\theta) \subseteq \Delta$ and let $w \in G$ be a representative for $\tilde{w}$. Let $\mathbf{N}_{\tilde{w}} = \mathbf{U} \cap w\mathbf{N}^{-}w^{-1}$, where $\mathbf{N}^{-}$ is the unipotent subgroup opposed to $\mathbf{N}$ (i.e. generated by negatives of roots in $\mathbf{N}$). Given $f \in V(\nu, \sigma)$, let

$$A(\nu, \sigma, w)f(g) = \int_{N_{\tilde{w}}} f(w^{-1}ng)dn \qquad (g \in G).$$

The integral converges if $\mathrm{Re}\langle \nu, H_\alpha \rangle$ is sufficiently large for each $\alpha \in \Delta - \theta$, where $H_\alpha \in \mathfrak{a}$ is the standard coroot attached to α; in particular $\langle \alpha, H_\alpha \rangle = 2$. Moreover, given f and g, $A(\nu, \sigma, w)f(g)$ extends to a meromorphic function of $\nu \in \mathfrak{a}^*_{\mathbf{C}}$ (cf. [39], [45]) and away from its poles, it intertwines representations $I(\nu, \sigma)$ and $I(\tilde{w}(\nu), \tilde{w}(\sigma))$ where $\tilde{w}(\sigma)(m') = \sigma(w^{-1}m'w)$, $m' \in M' = wMw^{-1}$. Finally, let $A(\sigma_\nu, w) = A(\nu, \sigma, w)$, where

$$\sigma_\nu = \sigma \otimes q^{\langle \nu, H_M(\ ) \rangle}.$$

It also satisfies $A(\sigma_\nu, w) = A(0, \sigma_\nu, w)$.

2 L-groups and L-functions

For a split group it is easy to define what its L-group is [3, 28]. Suppose $\mathbf{G}$ is a connected reductive split group over F. Fix a Cartan subgroup $\mathbf{T}$ of $\mathbf{G}$. There is a complex connected reductive group ${}^LG$, containing a Cartan subgroup ${}^LT$, whose Cartan matrix is the transpose of that of $\mathbf{G}$, and moreover $X(\mathbf{T}) = X_*({}^LT)$ and $X_*(\mathbf{T}) = X({}^LT)$. Here $X(\ )$ and $X_*(\ )$ denote the character and cocharacter (homomorphism of GL_1 into $\mathbf{T}$ or ${}^LT$) groups of the corresponding Cartan, respectively. The L-group ${}^LH$ is unique up to isomorphism.

Remark. In general ${}^LG$ has to be considered as a disconnected group ${}^LG^0 \rtimes W_{\bar{F}/F}$, where the Weil group $W_{\bar{F}/F}$ acts on ${}^LG^0$ through the Galois group (cf. [3], [48]). But since $\mathbf{G}$ is split this action is trivial and we may, as we in fact do, drop $W_{\bar{F}/F}$.

For a root α of $\mathbf{T}$, let $\alpha^\vee$ denote the corresponding coroot of $\mathbf{T}$ considered as a root of ${}^LT$ via $X_*(\mathbf{T}) = X({}^LT)$. Now, let $\mathbf{M}$ be a Levi subgroup in $\mathbf{G}$ generated by $\theta \subseteq \Delta$ (Δ is the set of simple roots of $\mathbf{T}$). Then ${}^LM$ is isomorphic to the Levi subgroup of ${}^LG$ generated by $\theta^\vee = \{\alpha^\vee \mid \alpha \in \theta\}$. Let ${}^LB$ be the Borel subgroup of ${}^LG$, containing ${}^LT$, and determined by $\Delta^\vee$. (If B is attached to Δ, then ${}^LB$ is the L-group of B.) There is a parabolic subgroup ${}^LP$ of ${}^LG$ which has ${}^LM$ as its Levi factor and ${}^LP \supseteq {}^LB$. Let ${}^LN$ be the unipotent radical of ${}^LP$. More generally, for any subgroup $\mathbf{R}$ of $\mathbf{U}$, the unipotent radical of $\mathbf{B}$, we use ${}^LR$ to denote the subgroup of ${}^LU$ generated by the coroots whose corresponding roots generate $\mathbf{R}$. Thus ${}^LN_{\tilde{w}}$ denotes the L-group of $\mathbf{N}_{\tilde{w}}$.

The group ${}^LM$ acts by adjoint action r on ${}^L\mathfrak{n}$, the Lie algebra of ${}^LN$. It leaves ${}^L\mathfrak{n}_{\tilde{w}}$, the Lie algebra of ${}^LN_{\tilde{w}}$, realized as a subspace of ${}^L\mathfrak{n}$ by $-\mathrm{Ad}(w)$, invariant. Let $r_{\tilde{w}}$ be the restriction of r on ${}^L\mathfrak{n}_{\tilde{w}}$.

Let $\rho = \rho_{\mathbf{P}}$ be half the sum of positive roots generating $\mathbf{N}$, $\mathbf{P} = \mathbf{MN}$. If $(\ ,\ )$ is the standard inner product in $\mathbf{R}^\ell$, $\ell = \mathrm{Card}(\Delta)$, let

$$\langle \alpha, \beta \rangle = 2\frac{(\alpha, \beta)}{(\beta, \beta)},$$

for every α and β in $\mathbf{R}^\ell$. Then for each α which has a root vector in $\mathbf{N}$, $\langle 2\rho, \alpha \rangle$ is a positive integer. Let $a_1 < a_2 < \cdots < a_m$ be distinct values of $\langle 2\rho, \alpha \rangle$. Set

$$V_i = \{X_{\alpha^\vee} \in {}^L\mathfrak{n}_{\tilde{w}} \mid \langle 2\rho, \alpha \rangle = a_i\}.$$

Each V_i is invariant under $r_{\tilde{w}}$. Let $r_{\tilde{w},i}$ be the restriction of $r_{\tilde{w}}$ to V_i (cf. [36]).

If $\mathbf{P}$ is maximal, i.e. $\Delta - \theta$ is singleton, then for the non-trivial $\tilde{w}$, we set $r_{\tilde{w}} = r$. Let $r_i = r_{\tilde{w},i}$ and thus $r = \bigoplus_{i=1}^m r_i$, with each r_i irreducible [40].

Moreover if $\alpha \in \Delta$ identifies the unique reduced root of **A** in **N**, we let $\tilde{\alpha} = \langle \rho, \alpha \rangle^{-1} \rho$, an element of $\mathfrak{a}^*$. Then for each i, $1 \leqslant i \leqslant m$,

$$V_i = \{X_{\beta^\vee} \in {}^L\mathfrak{n} \mid \langle \tilde{\alpha}, \beta \rangle = i\}.$$

The subgroup **T** is also a Cartan subgroup for **M**. Suppose σ has a vector fixed by $\mathbf{M}(O)$. Then there exists a complex character λ of T, trivial on $\mathbf{T}(O)$, which uniquely determines σ (cf. [3], [8]). It is in fact the unique constituent of $\mathrm{Ind}_{(B \cap M) \uparrow M} \lambda$ which has a vector fixed by $\mathbf{M}(O)$. Moreover if λ is changed to $\tilde{w}\lambda$, $\tilde{w} \in W(\mathbf{T}, \mathbf{M})$, σ remains unchanged. Identifying λ with an element $\lambda^\vee \in X(\mathbf{T}) \otimes_{\mathbf{Z}} \mathbf{C}^* = {}^LT$ now implies that σ is determined, uniquely, by the (semisimple) conjugacy class A of $\lambda^\vee$ in ${}^LM$.

Let ρ be a complex analytic (finite-dimensional) representation of the complex group ${}^LM$. If $s \in \mathbf{C}$, the local Langlands L-function attached to σ and ρ is defined as (cf. [3], [28]):

$$L(s, \sigma, \rho) = \det(I - \rho(A) q^{-s})^{-1}.$$

Suppose $f_0 \in V(\nu, \sigma)$ is fixed by $K = \mathbf{G}(O)$ and moreover $f_0(e) = 1$. Then it was proved by Langlands [26] that:

$$A(\nu, \sigma, w) f_0(e) = L(0, \sigma_\nu, r_{\tilde{w}}) / L(1, \sigma_\nu, r_{\tilde{w}}). \tag{2.1}$$

In particular if **P** is maximal:

$$A(s\tilde{\alpha}, \sigma, w_0) f_0(e) = \prod_{i=1}^{m} L(is, \sigma, r_i) / L(1 + is, \sigma, r_i). \tag{2.2}$$

We conclude this section with the following example.

Fix two positive integers m and n and let $\mathbf{G} = \mathrm{GL}_{n+m}$. We fix the subgroup of upper triangulars as our Borel subgroup. Let $\mathbf{M} = \mathrm{GL}_m \times \mathrm{GL}_n$ be the Levi subgroup of the standard maximal parabolic subgroup of **G** generated by

$$\theta = \{\alpha_1, \ldots, \alpha_{m-1}, \alpha_{m+1}, \ldots, \alpha_{m+n-1}\}.$$

Let σ_1 and σ_2 be two class 1 representations of $\mathrm{GL}_m(F)$ and $\mathrm{GL}_n(F)$, respectively. Then there exists unramified characters $\mu = (\mu_1, \ldots, \mu_m)$ and $\nu = (\nu_1, \ldots, \nu_n)$ of $(F^*)^m$ and $(F^*)^n$ such that σ_1 and σ_2 are constituents of $\mathrm{Ind}_{B_m(F) \uparrow \mathrm{GL}_m(F)} \mu \otimes \mathbf{1}$ and $\mathrm{Ind}_{B_n(F) \uparrow \mathrm{GL}_n(F)} \nu \otimes \mathbf{1}$, respectively. Here B_m and B_n are (Borel) subgroups of upper triangular matrices of GL_m and GL_n, respectively. Observe that $\mathbf{T}(F) = (F^*)^m \times (F^*)^n$.

Let $t_1 = \mathrm{diag}(\mu_1(\varpi), \ldots, \mu_m(\varpi))$ and $t_2 = \mathrm{diag}(\nu_1(\varpi), \ldots, \nu_n(\varpi))$ and let A_1 and A_2 be the corresponding (semisimple) conjugacy classes in

$\mathrm{GL}_m(\mathbb{C})$ and $\mathrm{GL}_n(\mathbb{C})$, respectively. If $\sigma = \sigma_1 \otimes \sigma_2$, then the semi-simple conjugacy class attached to σ is $A_1 \otimes A_2$ (Section 2). Denote by ρ_m and ρ_n standard representations of $\mathrm{GL}_m(\mathbb{C})$ and $\mathrm{GL}_n(\mathbb{C})$, respectively.

Consider $\mathbf{P} = \mathbf{MN}$, $\mathbf{M} = \mathrm{GL}_m \times \mathrm{GL}_n$, as a maximal parabolic subgroup of $\mathbf{G} = \mathrm{GL}_{n+m}$. Then in the notation of Section 2, $m = 1$ and $r = r_1 = \rho_m \otimes \tilde{\rho}_n$. The L-function $L(s, \sigma, r_1)$ is

$$\begin{aligned} L(s,\sigma,r_1) &= \det(I - (\rho_m \otimes \tilde{\rho}_n)(A_1 \otimes A_2)q^{-s})^{-1} \\ &= \prod_{i=1}^{m}\prod_{j=1}^{n}(1 - \mu_i(\varpi)\nu_j(\varpi)^{-1}q^{-s})^{-1}. \end{aligned}$$

In the notation of [16, 18, 38, 42] this is the Rankin–Selberg L-function $L(s, \sigma_1 \times \tilde{\sigma}_2)$ defined by Jacquet, Piatetski–Shapiro, and Shalika. We refer to their work [16, 18] as well as [38, 39, 41, 42, 32] for their global significance. We shall treat more examples in the last section.

Remark. We should remark that for the sake of simplicity, here we have chosen the inverse of the parametrization used in [26, 36, 40]. In fact if we use that, r_i in equations (2.1) and (2.2) must be changed to $\tilde{r}_i$ (equation (2.7) of [40]). In particular in our example t_1 and t_2 must be changed to t_1^{-1} and t_2^{-1} and the L-function that appears in equation (2.2) will be $L(s, \tilde{\sigma}_1 \times \sigma_2)$.

3 Plancherel measures

As is the case with abelian harmonic analysis, there is a certain measure which plays an important role in the Plancherel formula for G (cf. [14], [36]) which analogously is called the Plancherel measure. More precisely, for every conjugacy class of parabolic subgroups of G and an equivalence class $\{\sigma\}$ of discrete series representations of the corresponding Levi factors, there is a complex function $\mu(\nu,\sigma)$, $\nu \in \mathfrak{a}^*$, the so-called Plancherel measure, defined by

$$A(\nu,\sigma,w_0)A(\tilde{w}_0(\nu),\tilde{w}_0(\sigma),w_0^{-1}) = \mu(\nu,\sigma)^{-1}\gamma(G/P)\gamma(G/P').$$

Here $\tilde{w}_0 \in W(\mathbf{T})$ is the longest element modulo that of the Weyl group of $\mathbf{T}$ in $\mathbf{M}$ and

$$\gamma(G/P) = \int_{\mathbf{N}^-(F)} q^{\langle 2\rho_{\mathbf{P}}, H_P(n^-)\rangle}\, dn^-.$$

The measure dn^- is the transfer of the measure defining $A(\nu,\sigma,w_0)$ by means of $\mathbf{N}^- = w_0^{-1}\mathbf{N}_{\tilde{w}_0}w_0$. Similarly, we define $\gamma(G/P')$ for which $\mathbf{M}' = w_0\mathbf{M}w_0^{-1}$. It is easy to show that $\mu(\nu,\sigma)$ depends only on $\tilde{w}_0$ (and not w_0) and the class $\{\sigma\}$ of σ and is independent of the measures defining the intertwining operators.

Suppose σ is of class 1 (Section 2). Then by equation (2.1),

$$\mu(\nu,\sigma)^{-1}\gamma(G/P)\gamma(G/P') = L(0,\sigma_\nu,r)L(0,\sigma_\nu,\tilde{r})/L(1,\sigma_\nu,r)L(1,\sigma_\nu,\tilde{r}).$$

For a general σ and ρ, the L-function $L(s,\sigma,\rho)$ is not defined. But if $\rho = r$ (in fact for each $r_{\tilde{w},i}$) and σ is generic (see Section 5 below), the L-function $L(s,\sigma,r)$ and a monomial $\varepsilon(s,\sigma,r,\psi_F)$ in q^{-s}, depending on a non-trivial selfdual additive character ψ_F of F, the so-called root number attached to σ and r, are as defined in [36].

Chief among properties of these factors is that they are the local factors appearing in the functional equation satisfied by any globally generic (cf. [40]) cusp form which has σ as its local component (part 4 of Theorem 3.5 and Theorem 7.7 of [36]).

Now, let

$$\gamma(s,\sigma,r,\psi_F) = \varepsilon(s,\sigma,r,\psi_F)L(1-s,\tilde{\sigma},r)/L(s,\sigma,r).$$

Then Corollary 3.6 of [36] proves:

$$\mu(\nu,\sigma)\gamma(G/P)^{-1}\gamma(G/P')^{-1} = \gamma(0,\sigma_\nu,r,\psi_F)\gamma(0,\tilde{\sigma}_\nu,r,\tilde{\psi}_F). \tag{3.1}$$

This is equivalent to a conjecture of Langlands ([27], Appendix II) on normalization of intertwining operators by means of local factors which we shall return to in Section 7.

It is now clear that our knowledge of these local factors would lead to that of $\mu(\nu, \sigma)$. In the next section we shall discuss the importance of $\mu(\nu, \sigma)$ in reducibility questions for $I(\nu, \sigma)$.

4 R-groups

With notation as before, assume σ is a discrete series representation of $M = \mathbf{M}(F)$, i.e. its matrix coefficients are square integrable modulo the centre of M. Fix a positive root α which does not belong to $\mathbf{M} = \mathbf{M}_\theta$. Let $\mathbf{A}$ be the split torus in the centre of $\mathbf{M}$ and denote by $\mathbf{A}_\alpha$ the maximal subtorus in $\mathbf{A}$ which lies in the kernel of α. Let $\mathbf{M}_\alpha$ be the centralizer of $\mathbf{A}_\alpha$ in $\mathbf{G}$. The group $\mathbf{M}_\alpha \cap \mathbf{P}$ is a parabolic subgroup of $\mathbf{M}_\alpha$ with $\mathbf{M}$ as its Levi factor. Suppose $\tilde{w}_\alpha(\sigma) \cong \sigma$, where $\tilde{w}_\alpha$ is the reflection about α. Let $\mu_\alpha(\sigma)$ be the Plancherel measure attached to σ as a representation of $\mathbf{M}$ inside $\mathbf{M}_\alpha$. Define:

$$W(\sigma) = \{\tilde{w} \mid \tilde{w}(\sigma) \cong \sigma\}$$

and let $W''(\sigma)$ be the subgroup generated by those $\tilde{w}_\alpha$ in $W(\sigma)$ for which $\mu_\alpha(\sigma) = 0$. It is a normal subgroup of $W(\sigma)$. Set $R(\sigma) = W(\sigma)/W''(\sigma)$. It can be realized as a subgroup of $W(\sigma)$. The following theorem is due to Harish-Chandra [45] and Silberger [44]. We shall return to it in Section 7.

Theorem 4.1. (Harish-Chandra and Silberger). *The algebra of self-intertwining operators (commuting algebra) of $I(\sigma)$ is isomorphic to $\mathbb{C}[R(\sigma)]$. In particular the number of inequivalent components of $I(\sigma)$ is equal to the number of conjugacy classes in $R(\sigma)$.*

Therefore to understand the reducibility and composition factors for $I(\sigma)$ by means of R-groups, one needs to know Plancherel measures.

There is a product formula which reduces the calculation of Plancherel measures to those cases where $\mathbf{P}$ is maximal (Corollary 5.4.3.3 of [45]).

When $\mathbf{P} = \mathbf{B}$ and therefore σ is one-dimensional, the product formula reduces the problem to the well-known case of SL_2. While Plancherel measures are easy to compute, the problem of what possible R-groups are is non-trivial. In this case, R-groups have been studied and classified by Keys [19] (cf. [20] and [21] for some non-split groups). In particular, his work [19] implies the existence of non-abelian R-groups and therefore existence of constituents of $I(\sigma)$ which can appear with large multiplicities, an interesting feature of p-adic groups which does not appear for real groups. We refer to [24] for the original example of this phenomenon.

In conclusion, there are two problems that one must study, first to compute Plancherel measures for maximal parabolics and second to use such results to determine R-groups. It is the first problem that we shall address in the rest of this paper.

5 Generic and supercuspidal representations

From now on we shall assume that $\mathbf{P} = \mathbf{MN}$ is maximal. We recall that, given a pair of vectors v and $\tilde{v}$ in the spaces of σ and its contragredient $\tilde{\sigma}$, respectively, the function $g \mapsto \langle\sigma(g)v, \tilde{v}\rangle$ is called a matrix coefficient of σ. An irreducible representation σ of M is called supercuspidal if it has a non-zero matrix coefficient (and therefore all) which is of compact support modulo the centre of M.

Let ψ_F be a non-trivial additive character of F. For every $\alpha \in \Delta$, fix a vector X_α in the Lie algebra of U such that $\exp(tX_\alpha)$, $t \in F$, generates the one-parameter subgroup of G at α, i.e. fix the splitting (cf. [3]) for G. Define a character χ_0 of U by $\chi_0(\exp(tX_\alpha)) = \psi_F(t)$, $\alpha \in \Delta$. In the sense of [36, 40], χ_0 is clearly a *generic* character. Changing the splitting $\{X_\alpha \mid \alpha \in \Delta\}$ one obtains all the generic characters χ of U in this fashion. By restriction χ_0 is also a generic character of $U \cup M$.

An irreducible admissible representation σ of M is called χ-generic if it can be realized on a space of complex functions W on M, satisfying $W(um) = \chi(u)W(m)$, $u \in U \cap M$, $m \in M$. It is called generic if it is generic with respect to some generic character χ of $U \cap M$. We remark that, given a generic representation, one can change the splitting on G so that it becomes χ_0-generic. We refer to [33, 35, 36] for an interpretation of a generic representation in terms of its character.

If the simply connected covering of the derived group of $\mathbf{M}$ is a product of A-type groups, then every irreducible supercuspidal representation of M is generic with respect to some generic character of $U \cap M$. We refer to Conjecture 9.4 of [36] on how often these representations appear. Loosely speaking, they are expected to define all the root numbers and L-functions; in particular, those discussed in Section 2.

6 Plancherel measures for generic representations

Let $s \in \mathbb{C}$. Since $\mathbf{P}$ is maximal it is enough to compute $\mu(s\tilde{\alpha}, \sigma)$. Let r be the adjoint action of ${}^L M$ on ${}^L\mathfrak{n}$. Write $r = \bigoplus_i r_i$ with ordering as in Section 2. When σ is generic, in [36] we defined, for each i, $1 \leqslant i \leqslant m$, a unique root number $\varepsilon(s, \sigma, r_i, \psi_F)$ and a unique L-function $L(s, \sigma, r_i)$, satisfying a number of properties, chiefly to make them local factors appearing in the functional equation satisfied by every globally generic cusp form which has σ as its local component (Theorem 3.5 and Section 7 of [36]).

For each i, $1 \leqslant i \leqslant m$, let $P_{\sigma,i}(t) \in \mathbb{C}[t]$ be the polynomial satisfying

$$L(s, \sigma, r_i) = P_{\sigma,i}(q^{-s})^{-1}.$$

In particular $P_{\sigma,i}(0) = 1$. Then putting together Lemma 7.5 and Corollary 7.6 of [36], we have:

Proposition 6.1. *Let σ be an irreducible supercuspidal generic representation of M. Then:*

i) For $i \geqslant 3$, each $P_{\sigma,i} \equiv 1$.

ii) The operator

$$P_{\sigma,1}(q^{-s})P_{\sigma,2}(q^{-2s})A(s\tilde{\alpha}, \sigma, \tilde{w}_0)$$

is a non-zero and holomorphic operator for all $s \in \mathbb{C}$.

iii) Up to a monomial in q^{-s}, $\mu(s\tilde{\alpha}, \sigma)$ is equal to

$$\frac{P_{\sigma,1}(q^{-s})P_{\sigma,2}(q^{-2s})P_{\tilde{\sigma},1}(q^{s})P_{\tilde{\sigma},2}(q^{2s})}{P_{\sigma,1}(q^{-(1+s)})P_{\sigma,2}(q^{-(1+2s)})P_{\tilde{\sigma},1}(q^{-(1-s)})P_{\tilde{\sigma},2}(q^{-(1-2s)})}$$

in which the numerator and the denominator have no common factors. Consequently, if moreover σ is unitary, then the following statements are equivalent:

(a) σ is a pole of $A(\sigma, w_0)$.

(b) $P_{\sigma,i}(1) = 0$ for either $i = 1$ or 2 and only for one of them.

(c) $I(\sigma)$ is irreducible and σ is ramified, i.e. $\tilde{w}_0(\sigma) \cong \sigma$.

It is easy to check the equivalence of (b) and (c) if we are willing to take the discussion on R-groups for granted. By part 3 it is clear that $\mu(0, \sigma) = 0$ if and only if either $P_{\sigma,1}(0) = 0$ or $P_{\sigma,2}(0) = 0$. It is well known that since σ is supercuspidal and $\mathbf{P}$ is maximal $W(\sigma)$ has at most two elements. But if $\mu(0, \sigma) = 0$, then $W''(\sigma) \subseteq W(\sigma)$ has $\tilde{w}_0$ as its non-trivial element. Thus $W''(\sigma) = W(\sigma) \cong \mathbb{Z}/2\mathbb{Z}$. Consequently, $R(\sigma)$ is trivial and $I(\sigma)$ is irreducible. Conversely (c) implies that $R(\sigma)$ is trivial and $W(\sigma) = \mathbb{Z}/2\mathbb{Z}$. Then $W''(\sigma) = \mathbb{Z}/2\mathbb{Z}$, i.e. $\tilde{w}_0 \in W''(\sigma)$. As a result $\mu(0, \sigma) = 0$ which implies (b).

The fact that only one of the two polynomials $P_{\sigma,1}$ and $P_{\sigma,2}$ can vanish at 1 follows from part 2 and the simplicity of the poles of $A(s\tilde{\alpha}, \sigma, \tilde{w}_0)$ for maximal $\mathbf{P}$ and supercuspidal σ (cf. [45] for c-functions together with [39]).

Remark 1. The local coefficient $C_\chi(s\tilde{\alpha}, \sigma, w_0)$ in the statement of Corollary 7.6 of [36] must in fact be $C_{\bar{\chi}}(s\tilde{\alpha}, \tilde{\sigma}, w_0)$, since in [36] our parametrization is different from the one here (cf. the remark at the end of Section 2).

Remark 2. Since the Plancherel measure is independent of the choice of splitting, the results are independent of the choice of χ with respect to which σ is generic.

Remark 3. Accepting Conjectures 9.2 and 9.4 of [36], one can compute Plancherel measures even for non-generic representations. For this we refer to Theorem 9.5 of [36].

7 Complementary series and normalization of intertwining operators

When **M** is maximal and σ is supercuspidal and generic, the polynomials $P_{\sigma,i}$, $i = 1,2$, not only determine the reducibility of $I(s\tilde{\alpha},\sigma)$ on the imaginary axis but also off it. In fact, the knowledge of the $P_{\sigma,i}$, $i = 1,2$, provides us with complete understanding of complementary series in such cases. These are irreducible pre-unitary representations of the form $I(s\tilde{\alpha},\sigma)$ with $s > 0$. In fact Theorem 8.1 of [36] completely answers these questions and in particular shows that complementary series, if they exist, will only have $0 < s < 1/2$ or $0 < s < 1$. More precisely:

Theorem 7.1. *Let* $\mathbf{P} = \mathbf{MN}$ *be a maximal parabolic sugroup of* $\mathbf{G}$. *Fix an irreducible unitary generic supercuspidal representation* σ *of* M. *Assume* σ *is ramified, i.e.* $\tilde{w}_0(\sigma) \cong \sigma$, *and* $I(\sigma)$ *is irreducible. Choose (by Proposition 6.1 part (b)), a unique* i, $i = 1,2$, *such that* $P_{\sigma,i}(1) = 0$. *Then:*

(a) For $0 < s < 1/i$, *the representation* $I(s\tilde{\alpha},\sigma)$ *is irreducible and in the complementary series.*

(b) The representation $I(\tilde{\alpha}/i,\sigma)$ *is reducible with a unique generic subrepresentation which is in the discrete series. (More precisely its twist with a certain non-unitary character of* M *is in the discrete series.) It has a unique irreducible quotient (Langlands quotient) which is never generic. It is a pre-unitary non-tempered representation.*

(c) For $s > 1/i$, *the representations* $I(s\tilde{\alpha},\sigma)$ *are always irreducible and never in the complementary series.*

If σ *is ramified and* $I(\sigma)$ *is reducible, then no* $I(s\tilde{\alpha},\sigma)$, $s > 0$, *is pre-unitary. They are all irreducible.*

What is crucial in the proof is our knowledge of the poles of $\mu(s\tilde{\alpha},\sigma)$ for $s > 0$. In fact the edge of complementary series is exactly where the first pole of $\mu(s\tilde{\alpha},\sigma)$ lies for $s > 0$ (cf. [12]).

Moreover one needs to use normalized intertwining operators. They exist quite generally. For a general $\mathbf{P} = \mathbf{MN}$ and an irreducible unitary representation σ of M, these are scalar multiples of operators $A(\nu,\sigma,w)$, denoted by $\mathcal{A}(\nu,\sigma,w)$ satisfying:

(a) $\mathcal{A}(\nu,\sigma,w_1w_2) = \mathcal{A}(\tilde{w}_2(\nu),\tilde{w}_2(\sigma),w_1)\mathcal{A}(\nu,\sigma,w_2)$,

(b) $\mathcal{A}(\nu,\sigma,w)^* = \mathcal{A}(-\tilde{w}(\bar{\nu}),\tilde{w}(\sigma),w^{-1})$,

i.e. $\mathcal{A}(\nu,\sigma,w)$ is unitary if $\nu = -\bar{\nu}$. Here adjoint is computed on each K-type.

The isomorphism in Theorem 4.1 is in fact induced from the map $\tilde{w} \mapsto \sigma(w)\mathcal{A}(\sigma,w)$, from $W(\sigma)$ into the commuting algebra of $I(\sigma)$ (cf.

Lemma 7.9 and Section 13 of [22] for the definition of $\sigma(w)$). Observe that $\sigma(w)\mathcal{A}(\sigma, w)$ is independent of the choice of w.

By condition (a) here and Lemma 13.1 of [22] this map is a projective homomorphism which can be made a homomorphism since $\mathbf{G}$ is (quasi)-split and σ is generic. This last remark was observed by Keys [21]. Finally we remark that this map is not even a projective homomorphism if one uses the unnormalized operators $A(\sigma, w)$. The subgroup $W''(\sigma)$ of $W(\sigma)$ defined in Section 4 is exactly the set of those $\tilde{w}$ for which $\sigma(w)\mathcal{A}(\sigma, w)$ acts as a scalar [44]. In other words, it can be shown that $W''(\sigma)$ is the kernel of the map $\tilde{w} \mapsto \sigma(w)\mathcal{A}(\sigma, w)$ when one passes to the projective group [44].

Langlands' conjecture discussed in Section 3 demands that such normalizations must be possible by means of local root numbers and L-functions attached to r (Section 2). In fact (a) and (b) must hold if

$$\mathcal{A}(\nu, \sigma, w) = \varepsilon(0, \sigma_\nu, r_{\tilde{w}}, \psi_F) L(1, \sigma_\nu, r_{\tilde{w}}) L(0, \sigma_\nu, r_{\tilde{w}})^{-1} A(\nu, \sigma, w).$$

For a generic σ, this is Theorem 7.9 of [36]. Assuming Conjectures 9.2 and 9.4 of [36], the general case is Theorem 9.5 of that paper.

We should remark that if one does not insist on such fine normalizing factors, then the existence of normalized operators is due to Langlands (cf. [50]). We refer to [1] for other expected properties of normalized operators.

8 Complementary series for Sp_{2n}

If $\mathbf{G}$ is split of semisimple rank 1, then the simply connected covering of its derived group is always SL_2 and therefore its representation theory is basically that of $SL_2(F)$. In this case Plancherel measures are well known (cf. [20]) and the complementary series are classical.

On the other hand, when $\mathbf{G}$ is split of semisimple rank 2 and $\mathbf{M}$ is maximal, polynomials $P_{\sigma,1}$ and $P_{\sigma,2}$ are accessible (a supercuspidal σ is automatically generic in these cases). Consequently in [36] and [43], Theorem 7.1 was successfully applied to determine completely all the complementary series and non-tempered representations coming from maximal parabolic subgroups of split rank 2 groups when the inducing representations are supercuspidal. In these cases polynomials $P_{\sigma,1}$ and $P_{\sigma,2}$ were determined through their definition as inverses of local L-functions $L(s,\sigma,r_1)$ and $L(s,\sigma,r_2)$. For example, when $\mathbf{G}$ is the exceptional split group of type G_2 and $\mathbf{M} = GL_2$ is generated by its long root,

$$r_1 = \mathrm{Sym}^3(\rho_2) \otimes (\Lambda^2\rho_2)^{-1},$$

while $r_2 = \Lambda^2\rho_2$. Here ρ_2 is the standard representation of ${}^LM = GL_2(\mathbb{C})$. The representation $\mathrm{Sym}^3(\rho_2)$ is the four-dimensional irreducible representation of $GL_2(\mathbb{C})$ on the space of symmetric tensors of rank 3, and $\Lambda^2\rho_2$ is the one-dimensional exterior square representation of $GL_2(\mathbb{C})$; i.e. $\Lambda^2\rho_2(g) = \det g$, $g \in GL_2(\mathbb{C})$ (cf. [36], [37]). If ω is the central character of σ, then

$$L(s,\sigma,\Lambda^2\rho_2) = L(s,\omega),$$

the Hecke L-function attached to ω. On the other hand, the L-function $L(s,\sigma,r_1)$ has been completely calculated in [37] (cf. Proposition 1.1 of [37]). The results are recorded in Proposition 8.3 of [36].

When $\mathbf{G} = GL_n$ or SL_n, Plancherel measures are calculated quite explicitly in [43] (based on the results of [38] and [16]). Again for a maximal parabolic $\mathbf{P} = \mathbf{MN}$ and supercuspidal σ, Theorem 7.1 applies and the results duplicate those of Bernstein–Zelevinski [2, 49]. We finally refer to the example at the end of Section 2 concerning the type of L-functions which appear.

In this section we shall state a theorem on complementary series for $Sp_{2n}(F)$, the symplectic group in $2n$-variables (of rank n), coming from supercuspidal representations of its GL_n Levi subgroup, i.e. the one generated by its short simple roots. The proof will appear elsewhere.

In this case $r_1 = \rho_n$ and $r_2 = \Lambda^2\rho_n$, where ρ_n is the standard representation of $GL_n(\mathbb{C})$, the L-group of $\mathbf{M} = GL_n$. The exterior square representation $\Lambda^2\rho_n$ of $GL_n(\mathbb{C})$ on alternating tensors of rank 2 is irreducible of dimension $\frac{1}{2}n(n-1)$.

While the identity (for the supercuspidal representation σ)

$$L(s,\sigma,\rho_n)=1$$

is well known [10], the L-function $L(s,\sigma,\Lambda^2\rho_n)$ has not yet been understood [6, 17]. One then has to compute the polynomial $P_{\sigma,2}$ using part 2 of Proposition 6.1. In fact, in general, this is how one can hope to compute these polynomials since the theory of automorphic L-functions developed from other methods (integral representations, for example), although useful for low rank groups and GL_n, must be completely understood (even in the known cases), if one wants to apply it to our theory.

We recall that

$$\mathrm{Sp}_{2n}(F)=\{g\in \mathrm{GL}_{2n}(F)\mid {}^tgJg=J\},$$

where

$$J=\begin{pmatrix}0 & I_n\\ -I_n & 0\end{pmatrix}\in \mathrm{GL}_{2n}(F).$$

Let

$$M=\left\{\begin{pmatrix}g & 0\\ 0 & {}^tg^{-1}\end{pmatrix}\mid\ g\in \mathrm{GL}_n(F)\right\}.$$

Then $P=MN$ is the standard maximal parabolic subgroup of $\mathrm{Sp}_{2n}(F)$ whose Levi factor M is the F-points of the Levi subgroup of Sp_{2n} generated by short simple roots of Sp_{2n}. Here

$$N=\left\{\begin{pmatrix}I_n & X\\ 0 & I_n\end{pmatrix}\mid\ X\in M_n(F),\ {}^tX=X\right\}.$$

Let σ be an irreducible unitary supercuspidal representation of $\mathrm{GL}_n(F)$. It is automatically generic. Let $I(s,\sigma)=I(s\tilde{\alpha},\sigma)$. Then

$$I(s,\sigma)=\operatorname*{Ind}_{MN\uparrow G}(\sigma\otimes|\det(\ )|^s)\otimes\mathbf{1}.$$

Set $I(\sigma)=I(0,\sigma)$.

If φ is a matrix coefficient of σ, then there exists a function $f\in C_c^\infty(\mathrm{GL}_n(F))$ such that

$$\varphi(g)=\int\limits_{Z_n(F)} f(zg)\omega^{-1}(z)dz, \tag{8.1}$$

where $Z_n(F)\cong F^*$ is the centre of $\mathrm{GL}_n(F)$ and ω is the central character of σ. We observe that σ being ramified (as a representation of M) immediately implies $\sigma\cong\tilde{\sigma}$. Finally, let

$$w = \begin{pmatrix} 0 & & & 1 \\ & & -1 & \\ & 1 & & \\ \cdot^{\cdot^{\cdot}} & & & 0 \end{pmatrix} \in \mathrm{GL}_n$$

and define the automorphism θ of GL_n by $\theta(g) = w^{-1t}g^{-1}w$. The following theorem can be proved.

Theorem 8.1.

(a) Suppose $n > 1$ is odd. Then $I(\sigma)$ is reducible if and only if $\sigma \cong \tilde{\sigma}$. In this case there is no reducibility off the unitary axis and no complementary series.

(b) Suppose n is even and $\sigma \cong \tilde{\sigma}$. Then $I(\sigma)$ is irreducible if and only if $\omega = 1$ and there exists a function $f \in C_c^\infty(\mathrm{GL}_n(F))$ for which φ defined by equation 8.1 is a matrix coefficient of σ, such that

$$\sum_\alpha \int_{G_{\theta,w^{-1}\alpha}(F)\backslash \mathrm{GL}_n(F)} f(\theta(g)w^{-1}\alpha g^{-1}) \neq 0. \qquad (8.2)$$

Here

$$G_{\theta,w^{-1}\alpha} = \left\{ g \in \mathrm{GL}_n \;\middle|\; \theta(g)w^{-1}\alpha g^{-1} = w^{-1}\alpha \right\},$$

the twisted centralizer of $w^{-1}\alpha$, and the sum runs over diagonal matrices α each defining an inequivalent class of quadratic forms over F. The points of reducibility for $I(s,\sigma)$ are then at $s = \pm 1/2$. Otherwise (still assuming $\sigma \cong \tilde{\sigma}$), $I(\sigma)$ is reducible and there are no complementary series.

Part (a) is proved directly using SO_{2n}. In particular the non-vanishing condition of equation (8.2) is never possible if $n > 1$ is odd.

The proof of part (b) is by direct calculation of $P_{\sigma,2}$, using part 2 of Proposition 6.1.

What is interesting about equation (8.2) is that it is a sum of θ-twisted orbital integrals of θ-conjugacy classes within several stable singular θ-twisted conjugacy classes, and its non-vanishing seems to be interpretable in terms of the theory of twisted endoscopy which is being developed by Kottwitz and Shelstad.

In view of parametrization problems [3] for representations of $\mathrm{GL}_n(F)$ by means of homomorphisms of the Deligne–Weil group into $\mathrm{GL}_n(\mathbb{C})$, this is as close as one can get to answer reducibility questions by means of R-groups (cf. the introduction of [17]). We hope to discuss these matters in a future paper, where we shall be giving the proof of Theorem 8.1 as well.

References

1. J. Arthur, 'Intertwining operators and residues I, weighted characters', J. Funct. Anal. **84** (1989), 19–84.

2. I. N. Bernstein and A. V. Zelevinsky, 'Induced representations of reductive p-adic groups I', Ann. Sci. Éc. Norm. Super. **10** (1977), 441–472.

3. A. Borel, 'Automorphic L-functions', Proc. Sympos. Pure Math., AMS, **33, II** (1979), 27–61.

4. A. Borel, 'Linear algebraic groups', Mathematics Lecture Note Series, W. A. Benjamin, Amsterdam, New York (1969).

5. A. Borel, 'Linear algebraic groups', Graduate Texts in Mathematics **126**, Springer-Verlag, New York, Heidelberg, Berlin, to be published.

6. D. Bump and S. Friedberg, 'The exterior square automorphic L-functions on GL(n)', in Festschrift in honor of I. I. Piatetski-Shapiro, Part II, Israel Mathematical Conference Proceedings, **3**, (1990), 47–65.

7. C. Bushnell, 'Hereditary orders, Gauss sums and supercuspidal representations of GL(N)', J. reine angew. Math. **375/376** (1987), 184–210.

8. P.Cartier, 'Representations of p-adic groups', Proc. Symp. Pure Math., AMS, **33**, I (1979), 111–155.

9. W.Casselman, 'Some general results in the theory of admissible representations of p-adic reductive groups', preprint.

10. R. Godement and H. Jacquet, 'Zeta functions of simple algebras', Lecture Notes in Math. **260**, Springer-Verlag, Berlin–Heidelberg–New York (1972).

11. Harish-Chandra, 'Collected Papers', **IV**, Springer-Verlag, Berlin-Heidelberg-New York (1984).

12. Harish-Chandra, 'On the theory of Eisenstein integrals', Conference in Harmonic Analysis, College Park, Maryland, Lecture Notes in Math., **266**, Springer-Verlag, Berlin–Heidelberg–New York (1972), 123–149; also in [11], 47–73.

13. Harish-Chandra, 'Harmonic analysis on reductive p-adic groups', Proc. Symp. Pure Math., AMS, **26** (1973), 167–192; also in [11], 75–100.

14. Harish-Chandra, 'The Plancherel formula for reductive p-adic groups', in Collected Papers IV, Springer-Verlag, Berlin–Heidelberg–New York (1984), 353–367.

15. R. Howe and A. Moy, 'Hecke algebra isomorphisms for GL_n over a p-adic field', J. Algebra (to appear).

16. H. Jacquet, I. I. Piatetski-Shapiro and J. A. Shalika, 'Rankin–Selberg convolutions', Am. J. Math. **105** (1983), 367–464.

17. H. Jacquet and J. A. Shalika, 'Exterior square L-functions', in Proceedings of the Conference on Automorphic Forms, Shimura Varieties, and L-functions, (ed. L. Clozel and J. S. Milne) in Perspectives in Mathematics, **11**, Academic Press (1990), 143–226.

18. H. Jacquet and J.A. Shalika, 'On Euler products and the classification of automorphic representations I, II', Amer. J. Math. **103** (1981), 499–558 and 777–815.

19. D. Keys, 'On the decomposition of reducible principal series representation of p-adic Chevalley groups', Pacific J. Math. **101** (1982), 351–388.

20. D. Keys, 'Principal series representations of special unitary groups over a local field', Compos. Math. **51** (1984), 115–130.

21. D. Keys, 'L-indistinguishability and R-groups for quasi-split groups: Unitary groups of even dimension', Ann. Sci. Éc. Norm. Super. **20** (1987), 31–64.

22. A. W. Knapp and E. M. Stein, 'Intertwining operators for semisimple groups, II', Invent. Math. **60** (1980), 9–84.

23. A. W. Knapp and G. Zuckerman, 'Normalizing factors, tempered representations, and L-groups', Proc. Symp. Pure Math., AMS, **33**, I (1979), 93–105.

24. A. W. Knapp and G. Zuckerman, 'Multiplicity one fails for p-adic unitary principal series', Hiroshima Math. J. **10** (1980), 295–309.

25. P. Kutzko, 'Towards a classification of supercuspidal representations of GL_N', J. London Math. Soc. **37** (1988), 265–274.

26. R. P. Langlands, 'Euler products', Yale University Press, New Haven, CT (1971).

27. R. P. Langlands, 'On the functional equations satisfied by Eisenstein series', Lecture Notes in Math. **544**, Springer-Verlag, Berlin–Heidelberg–New York (1976).

28. R. P. Langlands, 'Problems in the theory of automorphic forms', in Lecture Notes in Math. **170**, Springer-Verlag, Berlin–Heidelberg–New York, 18–86.

29. R. P. Langlands, 'On the classification of irreducible representations of real algebraic groups', in Representation theory and harmonic analysis on semisimple Lie groups, (ed. P. J. Sally, Jr. and D. A. Vogan), Mathematical Surveys and Monographs, AMS, **31** (1989), 101–170.

30. R. P. Langlands, 'Representation Theory: its rise and its role in number theory', Proceedings of the Gibbs Symposium, AMS, (1990), 181–210.

31. G. Lusztig, 'Characters of reductive groups over a finite field', Annals of Math. Studies **107**, Princeton University Press, Princeton, NJ.

32. C. Moeglin and J.-L. Waldspurger, 'Le spectre residual de GL(n)', Ann. Sci. Éc. Norm. Super. **22** (1989), 605–674.

33. C. Moeglin and J.-L.Waldspurger, 'Modèles Whittaker dégénérés pour des groupes p-adiques', Math. Z. **196** (1987), 427–452.

34. L. Morris, 'Tamely ramified supercuspidal representations of classical groups I, II', preprint.

35. F. Rodier, 'Modèle de Whittaker et caractères de reprèsentations', in Non-Commutative Harmonic Analysis, Lecture Notes in Math. **466**, Springer-Verlag, Berlin–Heidelberg–New York (1975), 151–171.

36. F. Shahidi, 'A proof of Langlands conjecture on Plancherel measures; complementary series for p-adic groups', Ann. Math. **132** (1990), 273–330.

37. F. Shahidi, 'Third symmetric power L-functions for GL(2)', Compos. Math. **70** (1989), 245–273.

38. F. Shahidi, 'Fourier transforms of intertwining operators and Plancherel measures for GL(n)', Amer. J. Math. **106** (1984), 67–111.

39. F. Shahidi, 'On certain L-functions', Amer. J. Math. **103** (1981), 297–356.

40. F. Shahidi, 'On the Ramanujan conjecture and finiteness of poles for certain L-functions', Ann. Math. **127** (1988), 547–584.

41. F. Shahidi, 'Local coefficients as Artin factors for real groups', Duke Math. J. **52** (1985), pp. 973–1007.

42. F. Shahidi, 'Automorphic L-functions: a survey', in Proceedings of the Conference on Automorphic Forms, Shimura Varieties, and L-functions, (ed. L. Clozel and J. S. Milne) in Perspectives in Mathematics, **10**, Academic Press (1990), 415–437.

43. F. Shahidi, 'Langlands' conjecture on Plancherel measures for p-adic groups', in Proceedings of the Conference on Harmonic Analysis on Reductive Groups, Bowdoin College, August 1989, to appear.

44. A. Silberger, 'The Knapp–Stein dimension theorem for p-adic groups', Proc. Amer. Math. Soc. **68** (1978), 243–246.

45. A. Silberger, 'Introduction to Harmonic Analysis on Reductive p-adic groups', Math. Notes of Princeton University Press, **23**, Princeton, NJ (1979).

46. R. Steinberg, 'Lectures on Chevalley groups', Yale University, 1968.

47. M. Tadic, 'On Jacquet modules of induced representations of p-adic symplectic groups', preprint.

48. J. Tate, 'Number theoretic background', Proc. Symp. Pure Math., AMS, **33**, II (1979), 3–26.

49. A. V. Zelevinsky, 'Induced representations of reductive p-adic groups II, on irreducible representations of $\mathrm{GL}(n)$', Ann. Sci. Éc. Norm. Super. **13** (1980), 165–210.

50. L. Clozel, J.-P. Labesse and R. P. Langlands, 'Morning seminar on the trace formula', Lecture Notes, Institute for Advanced Study, Princeton University.

51. L. Corwin, 'Representations of division algebras over local fields', Adv. Math. **13** (1974), 259–267.

8.

IWASAWA THEORY, FACTORIZABILITY AND THE GALOIS MODULE STRUCTURE OF UNITS

DAVID R. SOLOMON

UFR de Mathématiques,
Université de Bordeaux I,
351 Cours de la Libération,
33405 Talence Cedex,
France.

Introduction

The aim of this paper is to present a new, p-adic approach to a certain problem concerning the structure of algebraic number fields. The problem is that of describing the group of units (more generally the S-units) of an extension field as a module for the Galois group. Our approach involves the application of Iwasawa's theory of $\mathbb{Z}_p$-extensions.

The main module-theoretic tool will be the concept of *factor equivalence*. This is an equivalence relation between lattices acted upon by a finite abelian group G and is a weakening of (local) isomorphism as G-modules. The idea was first introduced in [19]. Much of the fundamental work is due to Fröhlich and throughout its development the emphasis has been on applications to questions of Galois module structure (see [19] and also, for example, [3]). Factor equivalence was first applied to the study of units in [10]. This paper, together with [9], are the main references for Sections 1 and 3 in which the basic ideas and results of factorizability and factor equivalence are explained. They should be consulted for further details and proofs of some results. (Our notation will differ somewhat from theirs however. Note in particular that our 'weakly factorizable' is their 'factorisable' and what we call 'factorizable' would be '$\mathbb{Q}_p$-factorisable' in their notation.)

Fröhlich's factor equivalence results on units were obtained by means

of complex Artin L-functions and a new type of regulator of units. By employing instead ideas from Iwasawa Theory we obtain results of a similar nature, whose proof is, however, wholly independent and algebraic. In fact, in certain cases these results go further than those of [10] (for example, for totally real fields abelian over the rational numbers). This is the content of Sections 2 and 4. Section 2 includes some of the Iwasawa-theoretic background. For a more complete introduction to this area see for example [23] or [16]. Section 4 contains the main result of the paper and a sketch of its proof. Some of the ideas are similar to ones that have been used by Coates (see Theorem 1.13 and the Appendix to [5]) and to those used by Greenberg in [12] and Gillard in [11] in their deductions of the conjecture of Gras on the cyclotomic units from the so-called Main Conjecture of Iwasawa Theory over $\mathbb{Q}$. We make no use of the latter conjecture, however. (It was, of course, established as a theorem by Mazur and Wiles in [18].) Proofs of our results in greater detail and generality will appear in [20] in the context of the so-called 'canonical' factor equivalence of S-units.
Acknowledgement: The author was supported by a grant from the SERC whilst this work was carried out.

0.1 Notations, conventions, definitions

We write $\mathbb{N}$ for the natural numbers, $\mathbb{Z}$ for the ring of integers and $\mathbb{Q}$, $\mathbb{R}$ and $\mathbb{C}$ for the fields of rational, real and complex numbers respectively. Given a prime number $p \in \mathbb{Z}$, we shall denote by $\mathbb{Z}_p$ and $\mathbb{Q}_p$ the p-adic integers and the p-adic rational numbers respectively. We fix once and for all algebraic closures $\bar{\mathbb{Q}}$ of $\mathbb{Q}$ and $\bar{\mathbb{Q}}_p$ of $\mathbb{Q}_p$ for each prime p.

Given a number field $F \subseteq \bar{\mathbb{Q}}$, $\mathcal{O}_F$ will denote the algebraic integers of F and $W(F)$ the group of all roots of unity lying in F. The group of units and the class-group of $\mathcal{O}_F$ will be denoted respectively $E(F)$ and $\mathrm{Cl}(F)$. More generally, suppose that we are given a finite set T of finite places of F, identified with the corresponding prime ideals of $\mathcal{O}_F$. Then we shall write $E_T(F)$ and $\mathrm{Cl}_T(F)$ respectively for the units and class-group of the T-*integers* of F. (Explicitly, $E_T(F)$ is the subgroup of $F^\times$ consisting of those elements which are units of the localization of $\mathcal{O}_F$ at each prime outside T. $\mathrm{Cl}_T(F)$ is the quotient of $\mathrm{Cl}(F)$ by the subgroup generated by the classes of the ideals in T.) $w(F)$ will denote $|W(F)|$, $h(F)$ will denote $|\mathrm{Cl}(F)|$ and $h_T(F)$ will denote $|\mathrm{Cl}_T(F)|$. For a positive integer n, $F(n)$ will denote the field obtained by adjoining all nth roots of unity to F.

If R is a ring and G a finite group then $R[G]$ (or simply RG) will denote the group ring with coefficients in R.

All modules will be left modules. If A is an RG-module then A^G and A_G will denote respectively the G-invariants and G-coinvariants of A. Thus A^G is the R-submodule of A consisting of all the elements fixed by G and A_G is the quotient R-module A/I_GA where I_G is the augmentation ideal of RG .

Finally, if $p \in \mathbb{Z}$ is a prime number then the process of localizing and completing at p will be denoted by a subscript p as shorthand. Thus if A (respectively f) is a $\mathbb{Z}$- or $\mathbb{Z}G$-module (respectively a homomorphism of such modules), then we shall frequently use 'A_p' (respectively 'f_p') to denote $A \otimes_{\mathbb{Z}} \mathbb{Z}_p$ (respectively $f \otimes_{\mathbb{Z}} \mathrm{id}_{\mathbb{Z}_p}$).

0.2 Arithmetic Galois lattices

We introduce the lattices of number-theoretic significance which constitute our main objects of study. Let $\mathcal{L}$ be a finite extension of $\mathbb{Q}$ or $\mathbb{Q}_p$ and let $\mathcal{R}$ denote its ring of integers. Let V be a finite-dimensional vector space over $\mathcal{L}$. A finitely generated $\mathcal{R}$-submodule $\mathfrak{A}$ of V will be called an ($\mathcal{R}$-)*lattice* if and only if the injection

$$\mathfrak{A} \otimes_{\mathcal{R}} \mathcal{L} \longrightarrow V$$

is an isomorphism (i.e. $\mathfrak{A}$ spans V over $\mathcal{L}$). If a finite group G acts on V we call $\mathfrak{A}$ an *$\mathcal{R}G$-lattice in V* if and only if it is G-stable. A given, finitely generated, $\mathcal{R}$-torsion-free $\mathcal{R}G$-module $\mathfrak{A}$ will be regarded as an $\mathcal{R}G$-lattice within $\mathfrak{A} \otimes_{\mathcal{R}} \mathcal{L}$. If $\mathfrak{B}$ is another such, we shall say that $\mathfrak{A}$ and $\mathfrak{B}$ are *isogenous* if and only if $\mathfrak{A}\otimes_{\mathcal{R}}\mathcal{L}$ and $\mathfrak{B}\otimes_{\mathcal{R}}\mathcal{L}$ are isomorphic as $\mathcal{L}G$-modules. (This is a *non*-standard but convenient piece of terminology.) Of course, isogeny is in general a much weaker equivalence relation between lattices than isomorphism as $\mathcal{R}G$-modules. In fact, questions of Galois module structure frequently take the form of comparing two lattices which are isogenous as modules for a Galois group and asking when they are isomorphic or locally isomorphic. For example, 'additive' questions frequently concern the ring of integers $\mathcal{O}_L$ of a number field L. Suppose that L is a Galois extension of K with group G and let n denote the degree $[K : \mathbb{Q}]$. Then $\mathcal{O}_L$ may be regarded as an $\mathcal{O}_K G$-lattice (isogenous to $\mathcal{O}_K G$ itself, by the Normal Basis Theorem) or, by restriction of scalars, as a $\mathbb{Z}G$-lattice isogenous to $(\mathbb{Z}G)^n$. On the other hand, $\mathcal{O}_L$ is not in general even locally isomorphic to $(\mathbb{Z}G)^n$, (i.e. isomorphic after tensoring with $\mathbb{Z}_p$ for each p). In fact, we have:

Theorem 0.1. *$\mathcal{O}_{L,p}$ is isomorphic to $(\mathbb{Z}_p G)^n$ as $\mathbb{Z}_p G$-module if and only if all prime ideals of K above p are at most tamely ramified in L/K.*

Proof This is a consequence of a theorem of Noether; see for example [8], Theorem 3, p. 26. ◊ ◊

In the case where L/K is tamely ramified everywhere, $\mathcal{O}_L$ is (everywhere) locally free and through the work of Taylor, Fröhlich *et al.*, its global structure as $\mathbb{Z}G$-module is now very well understood (see [22] or the account in [8]). When L/K is wildly ramified our knowledge is less complete, even at the level of local isomorphism in general. Certain special cases *are* known, however (see for example [17] in the case where L is an abelian extension of $\mathbb{Q}$).

So-called 'multiplicative' questions concern the 'S-units'. Here S denotes a finite set of finite places of K and for each finite extension F of K we write $S(F)$ for the places of F above S and $E_S(F)$ for $E_{S(F)}(F)$ (the S-units). $E_S(L)$ is then a finitely generated $\mathbb{Z}G$-module with $\mathbb{Z}$-torsion submodule $W(L)$, hence the quotient

$$\mathfrak{U}_S(L) \stackrel{\text{def}}{=} E_S(L)/W(L)$$

is a $\mathbb{Z}G$-lattice. We shall compare this lattice with another, $\mathfrak{H}_S(L)$, defined as follows. Let $S_\infty(F)$ denote the infinite places of F; then G acts naturally on $S_\infty(L) \cup S(L)$ on the right and we write $\mathbb{Z}^{S_\infty(L)\cup S(L)}$ for the left $\mathbb{Z}G$-lattice

$$\mathrm{Hom}_{\mathbf{Set}}(S_\infty(L) \cup S(L), \mathbb{Z}).$$

$\mathfrak{H}_S(L)$ is then defined to be the kernel of the map

$$\begin{array}{ccc} \mathbb{Z}^{S_\infty(L)\cup S(L)} & \longrightarrow & \mathbb{Z} \\ f & \longmapsto & \sum_v f(v) \end{array}$$

and we have:

Proposition 0.2. *$\mathfrak{U}_S(L)$ and $\mathfrak{H}_S(L)$ are isogenous $\mathbb{Z}G$-lattices*

Proof See [21], Chapter 1, Section 4. ◊ ◊

Unfortunately, by comparison with the additive case, at present very little is known, even about the local structure of $\mathfrak{U}_S(L)$ or how it compares to that of $\mathfrak{H}_S(L)$. (One should mention here the pioneering work of Chinburg, however; see [4].) In this paper we shall follow Fröhlich in studying $\mathfrak{U}_S(L)$, $\mathfrak{H}_S(L)$ and related $\mathbb{Z}G$-lattices up to an equivalence relation stronger than isogeny although still weaker than local isomorphism. This is factor equivalence.

1 Factorizable Functions

1.1 Weak Factorizability

Let G be a finite abelian group. Fix an algebraically closed field $\mathbb{A}$ containing $\mathbb{Q}$ and write $\widehat{G}$ for the group of multiplicative characters χ of G taking values in $\mathbb{A}$. Say that two characters χ and χ' are *equivalent* if they generate the same cyclic subgroup of $\widehat{G}$. This is the same as saying that χ and χ' are 'conjugate over $\mathbb{Q}$', i.e. that there exists an automorphism ω of $\mathbb{A}$ with $\chi' = \omega \circ \chi =$ 'χ^{ω}'. A *division* D of G is then an equivalence class of characters and we write $\mathfrak{D}(G)$ for the set of divisions. Thus, for example, if C is cyclic of order p, a prime, then $|\mathfrak{D}(C)| = 2$ and $|\mathfrak{D}(C \times C)| = p+2$. Let $\mathcal{S}(G)$ denote the set of subgroups of G. A character is trivial on a subgroup $H \in \mathcal{S}(G)$ if and only if the same holds for all characters in the division D containing it and we then write $D(H) = 1$. Now let A be any abelian group, written multiplicatively.

Definition 1.1. *A function $f: \mathcal{S}(G) \to A$ is* weakly factorizable *if and only if there exists a function $h: \mathfrak{D}(G) \to A$ such that*

$$f(H) = \prod_{\substack{D \in \mathfrak{D}(G) \\ D(H)=1}} h(D)$$

REMARK 1. The Möbius inversion formula shows that if h exists it is unique and that if G is cyclic then every f is weakly factorizable

EXAMPLE 1. (See [1].) Let L/K be a Galois extension of number fields with abelian Galois group G. For each $H \in \mathcal{S}(G)$, we denote by $F = F(H)$ the fixed field L^H and identify G/H with $\mathrm{Gal}(F/K)$. For a complex variable s, $\zeta_F(s)$ denotes the Dedekind zeta function of F. Writing $\widetilde{\zeta_F}(0)$ for the leading term in its power series expansion at zero we have the classical 'class-number formula':

$$\widetilde{\zeta_F}(0) = -\frac{h(F)R_F}{w(F)} \tag{1.1.1}$$

where R_F is the regulator of units of F. For each complex-valued character ψ of G/H, we shall denote by $L_{F/K}(s,\psi)$ the L-function attached to ψ. Then there is a product formula:

$$\zeta_F(s) = \prod_{\psi \in \widehat{G/H}} L_{F/K}(s,\psi) \tag{1.1.2}$$

and finally there is the identity

$$L_{F/K}(s,\psi) = L_{L/K}(s,\mathrm{Infl}\psi) \qquad (1.1.3)$$

where $\mathrm{Infl}\psi$ denotes the element of $\widehat{G}$ obtained by inflating ψ to G. Taking $\mathcal{A} = \mathbb{C}$ and $A = \mathbb{C}^\times$ in the above, we have:

Proposition 1.2. *The function*

$$\begin{array}{rcl} f\colon \mathcal{S}(G) & \longrightarrow & \mathbb{C}^\times \\ H & \longmapsto & -\dfrac{h_F R_F}{w_F} \end{array}$$

is weakly factorizable.

Proof Let $\widetilde{L_{F/K}}(0,\psi)$ denote the leading term of $L_{F/K}(s,\psi)$ at $s = 0$. Then equations (1.1.1), (1.1.2) and (1.1.3) show that

$$f(H) = \prod_{\psi\in\widehat{G/H}} \widetilde{L_{L/K}}(0,\mathrm{Infl}\psi) = \prod_{\substack{\chi\in\widehat{G}\\ \chi(H)=1}} \widetilde{L_{L/K}}(0,\chi)$$

and thus in Definition 1.1 we may take

$$h(D) = \prod_{\chi\in D} \widetilde{L_{L/K}}(0,\chi).$$

◊ ◊

This example shows how arithmetic invariants can be factorized by means of L-functions.

1.2 Factorizability

We shall only need to consider this in a p-adic setting for a fixed prime p. For each finite extension k of $\mathbb{Q}_p$ within $\bar{\mathbb{Q}}_p$ we shall write $I(k)$ for the group of non-zero, fractional ideals of k under multiplication (this is isomorphic to $\mathbb{Z}$). If k is contained in k' there is a natural injection $I(k) \to I(k')$ obtained by extending ideals of k. Write $I(\bar{\mathbb{Q}}_p)$ for the direct limit over all k with respect to these maps (this is isomorphic to $\mathbb{Q}$). We regard each $I(k)$ as contained in $I(\bar{\mathbb{Q}}_p)$. Any element $a \in \bar{\mathbb{Q}}_p{}^\times$ generates an 'ideal' in $I(\bar{\mathbb{Q}}_p)$ written (a) or just a. We shall be considering functions of the form $f\colon \mathcal{S}(G) \to I(\bar{\mathbb{Q}}_p)$ which, taking A to be $I(\bar{\mathbb{Q}}_p)$ in the previous subsection, will be weakly factorizable. We shall also be dealing with $\bar{\mathbb{Q}}_p$-valued characters χ so we take $\mathcal{A}$ to be $\bar{\mathbb{Q}}_p$ and write $\mathbb{Q}_p(\chi)$ for the (cyclotomic) extension of $\mathbb{Q}_p$ generated by the values of χ. The following definition then strengthens the notion of weak factorizability for such a function f.

Definition 1.3. *We say that f is* factorizable *if and only if there exists a function $g: \widehat{G} \longrightarrow I(\bar{\mathbb{Q}}_p)$ satisfying:*

1. $g(\chi) \in I(\mathbb{Q}_p(\chi))$,
2. *for each* $\omega \in \mathrm{Gal}(\bar{\mathbb{Q}}_p/\mathbb{Q}_p)$, $g(\chi^\omega) = g(\chi)$,
3. *for each* $H \in \mathcal{S}(G)$:

$$f(H) = \prod_{\substack{\chi \in \widehat{G} \\ \chi(H)=1}} g(\chi).$$

REMARK 2. Since $\mathrm{Gal}(\bar{\mathbb{Q}}_p/\mathbb{Q}_p)$ acts trivially on $I(\bar{\mathbb{Q}}_p)$, condition 2 above 'really' says that g is Galois equivariant.

We recast this definition in an alternative form which is obviously equivalent as follows. Let Φ denote a $\mathbb{Q}_p$-conjugacy class of characters. Thus Φ is the set

$$\{\varphi = \chi^\omega : \omega \in \mathrm{Gal}(\bar{\mathbb{Q}}_p/\mathbb{Q}_p)\}$$

for some $\chi \in \widehat{G}$. Since $\mathbb{Q}_p(\varphi)$ depends only on Φ, we write it $\mathbb{Q}_p(\Phi)$. Similarly, we write $\Phi(H) = 1$ for $H \in \mathcal{S}(G)$ if $\varphi(H) = 1$ for one, hence any, $\varphi \in \Phi$ and we denote the set of all Φ by $X_p(G)$.

Alternative definition 1.3. *f is* factorizable *if and only if there exists a function*

$$g': X_p(G) \longrightarrow I(\bar{\mathbb{Q}}_p)$$

satisfying:

1'. $g'(\Phi)$ *lies in* $I(\mathbb{Q}_p)$,

2'. $g'(\Phi)$ *is the norm of an ideal of* $\mathbb{Q}_p(\Phi)$ *and*

3'. *for each* $H \in \mathcal{S}(G)$:

$$f(H) = \prod_{\substack{\Phi \in X_p(G) \\ \Phi(H)=1}} g'(\Phi).$$

REMARK 3.

(a) This shows in particular that any factorizable function f must take values in $I(\mathbb{Q}_p) \subseteq I(\bar{\mathbb{Q}}_p)$.

(b) Each division D is a union of conjugacy classes Φ so factorizability implies weak factorizability.

(c) Because one D may contain many of the φ however, g', if it exists, is not in general unique.

(d) Suppose that G is a p-group. Then divisions and $\mathbb{Q}_p$-conjugacy classes of characters coincide. Moreover, the norm maps are all surjective. It follows that any weakly factorizable f taking values in $I(\mathbb{Q}_p)$ will actually be factorizable and in this case $g' = h$ will be unique. To restore uniqueness to the factorization in the general case, one can decompose G as the product of its Sylow p-subgroup and a subgroup of order prime to p. This leads to an even stronger notion, that of *canonical factorizability*. (See [3].)

EXAMPLE 2. All constant functions with values in $I(\mathbb{Q}_p)$ are factorizable.

EXAMPLE 3. The product of two factorizable functions and the (pointwise) inverse of a factorizable function are again factorizable.

EXAMPLE 4. The following example is based on elementary facts concerning prime ideals in Galois extensions of number fields (see for example [15], Chapter 1, Section 5). Let L/K be such an extension having Galois group G. Let $\mathfrak{P} = \mathfrak{P}_K$ be a prime ideal of $\mathcal{O}_K$ lying above p. For each $H \in \mathcal{S}(G)$ we write F for the fixed field L^H and denote by g_F the number of prime ideals of $\mathcal{O}_F$ lying above $\mathfrak{P}$. These ideals are all Galois conjugate so if we choose one, say $\mathfrak{P}_F$, then they all have absolute norm equal to $N\mathfrak{P}_F \stackrel{\text{def}}{=} |\mathcal{O}_F : \mathfrak{P}_F|$. We have:

Proposition 1.4. *The function*

$$\begin{array}{rcl} c\colon \mathcal{S}(G) & \longrightarrow & I(\bar{\mathbb{Q}}_p); \\ H & \longmapsto & (N\mathfrak{P}_F)^{g_F} \end{array}$$

is factorizable.

Proof $N\mathfrak{P}_F$ is equal to $(N\mathfrak{P})^{f_F}$ where f_F is the degree of the residue field extension:

$$f_F = [\mathcal{O}_F/\mathfrak{P}_F : \mathcal{O}_K/\mathfrak{P}].$$

Now $f_F g_F$ is equal to the index in $\mathrm{Gal}(F/K)$ of the inertia group T_F of $\mathfrak{P}_F$. Moreover, under the identification of $\mathrm{Gal}(F/K)$ with G/H, T_F corresponds to TH/H where $T = T_L$ is the inertia group of $\mathfrak{P}_L$ in G. Thus we have:

$$\begin{array}{rcl} c(H) & = & (N\mathfrak{P})^{|G/H : TH/H|} \\ & = & (N\mathfrak{P})^{|G : TH|} \\ & = & (N\mathfrak{P})^{|\widehat{G/TH}|} \\ & = & (N\mathfrak{P})^{|\{\chi \in \widehat{G} : \chi(T)=1,\ \chi(H)=1\}|}. \end{array}$$

It follows that the function $g: \widehat{G} \to I(\mathbb{Q}_p)$ factorizes c where:

$$g(\chi) = \begin{cases} N\mathfrak{P} & \text{if } \chi(T) = 1 \quad (\chi \text{ 'unramified at } \mathfrak{P}\text{'}), \\ 1 & \text{if } \chi(T) \neq 1 \quad (\chi \text{ 'ramified at } \mathfrak{P}\text{'}). \end{cases}$$

◊ ◊

We leave the reader to check that, by contrast to these examples, the function $H \longmapsto |H|$ is not even weakly factorizable in general.

2 Iwasawa theory

In this section we shall see how Iwasawa's theory of $\mathbb{Z}_p$-extensions provides a source of factorizable functions. The results will be applied to questions of Galois structure in Section 4. We start by recalling some basic notations and results.

2.1 The cyclotomic $\mathbb{Z}_p$-extension

Let L/K be a finite, Galois extension of number fields inside $\bar{\mathbb{Q}}$. We shall assume in this section that L is *totally real*, i.e. every complex embedding of L actually maps it into $\mathbb{R} \subseteq \mathbb{C}$. Let p be a prime number fixed for this section and assumed for simplicity to be odd. For $n = 0, 1, 2, 3, \ldots$ the field $\mathbb{Q}(p^{n+1})$ has a unique subfield $\mathbb{B}_n$, cyclic of order p^n over $\mathbb{Q}$. We define $\mathbb{B}_\infty$ to be the increasing union:

$$\mathbb{B}_\infty = \bigcup_0^\infty \mathbb{B}_n \subseteq \bar{\mathbb{Q}}.$$

Then $\mathbb{B}_\infty/\mathbb{Q}$ is a $\mathbb{Z}_p$-extension, i.e. it is Galois with profinite Galois group isomorphic to $\mathbb{Z}_p$. For each subfield F of L containing K, we shall write F_∞ for the compositum $F\mathbb{B}_\infty \subseteq \bar{\mathbb{Q}}$. Thus F_∞/F is also a $\mathbb{Z}_p$-extension (the *cyclotomic* $\mathbb{Z}_p$-extension of F). It is unramified outside the set of primes of F dividing p. We also introduce:

$M(F) \stackrel{\text{def}}{=}$ the maximal abelian pro-p extension of F unramified outside the primes dividing p,

$M(F_\infty) \stackrel{\text{def}}{=}$ the maximal abelian pro-p extension of F_∞ unramified outside the primes dividing p.

Clearly there are inclusions $F_\infty \subseteq M(F) \subseteq M(F_\infty)$ as in the following diagram.

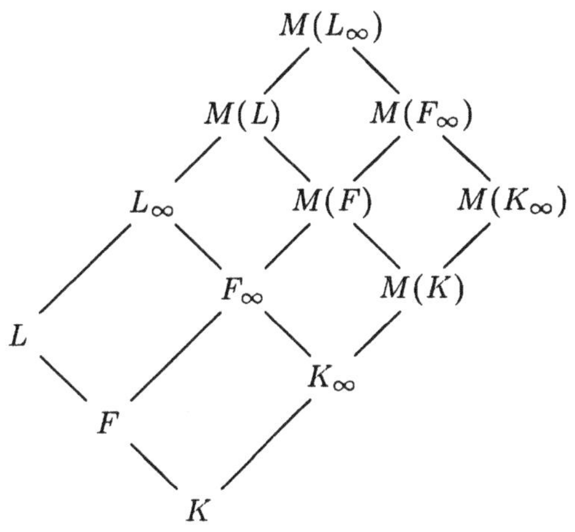

We shall write $\mathcal{X}(F)$ and $\mathcal{X}(F_\infty)$ for $\mathrm{Gal}(M(F)/F_\infty)$ and $\mathrm{Gal}(M(F_\infty)/F_\infty)$ respectively. We now introduce two hypotheses:

Hypothesis 2.1. *Leopoldt's conjecture is satisfied by L at p.*

This hypothesis, known to hold in certain cases (see [14] for example), is essential in what follows and will be assumed for the rest of this section. We recall briefly what it means. Since L is totally real, of degree r say, over $\mathbb{Q}$, we can choose a $\mathbb{Z}$-basis modulo torsion for $E(L)$ consisting of $r-1$ elements, $\epsilon_1, \ldots, \epsilon_{r-1}$. Leopoldt's conjecture is the statement that the p-adic regulator is non-zero, i.e.

$$\det(\log_p(\sigma_i(\epsilon_j)))_{i,j=1,\ldots,r-1} \neq 0$$

where $\log_p$ is the p-adic logarithm (see Section 4) and $\sigma_1, \ldots, \sigma_r$ are all the embeddings of K into $\bar{\mathbb{Q}}_p$ in some order (it does not matter which).

Proposition 2.2. *Under* Hypothesis 2.1, *$\mathcal{X}(F)$ is finite.*

Proof See [16], Chapter 5, Theorem 6.1. This is in fact equivalent to Hypothesis 2.1. ◊ ◊

Hypothesis 2.3. *K_∞ and L are linearly disjoint over K i.e. $K_\infty \cap L = K$.*

This hypothesis, which holds in the 'generic' case, is not at all necessary for the results of this section. Since it does greatly simplify the exposition however, we shall assume it henceforth. In particular, Hypothesis 2.3 allows us to identify, as we shall, the Galois groups $\mathrm{Gal}(F_\infty/F)$ as F varies and we shall denote these collectively by Γ. We also choose a topological generator γ of Γ, so that there is an isomorphism:

$$\begin{array}{ccc} \mathbb{Z}_p & \longrightarrow & \Gamma \\ a & \longmapsto & \gamma^a . \end{array}$$

2.2 Iwasawa polynomials

The extension $M(F_\infty)/F$ is Galois and pro-p although not in general abelian, and there is an exact sequence of groups:

$$0 \longrightarrow \mathcal{X}(F_\infty) \longrightarrow \mathrm{Gal}(M(F_\infty)/F) \longrightarrow \Gamma \longrightarrow 0.$$

This gives an action of Γ on $\mathcal{X}(F_\infty)$ (by extension and conjugation) and we have:

Proposition 2.4.

$$\mathcal{X}(F) \cong \mathrm{coker}(1-\gamma) \mid \mathcal{X}(F_\infty).$$

Proof See [16], Chapter 5, Section 6. ◇◇

In fact, the Γ-action extends to give $\mathcal{X}(F_\infty)$ the natural structure of a finitely generated topological module for the completed group-ring:

$$\mathbb{Z}_p[[\Gamma]] \stackrel{\text{def}}{=} \varprojlim_n \mathbb{Z}_p[\Gamma/\Gamma^{p^n}].$$

We may regard Γ as embedded in $\mathbb{Z}_p[[\Gamma]]^\times$ and it is well known that the choice of γ determines a topological isomorphism of compact local rings:

$$\begin{array}{ccc} \mathbb{Z}_p[[\Gamma]] & \cong & \mathbb{Z}_p[[T]] \quad \text{(power series)} \\ \gamma & \longleftrightarrow & 1+T. \end{array}$$

We denote $\mathbb{Z}_p[[T]]$ by Λ for brevity. The structure theorem for finitely generated Λ-modules now tells us: There exist non-negative integers r, s, t, positive integers λ_j, $j = 1, \ldots, t$, and μ_i, $i = 1, \ldots, s$, and polynomials f_j over $\mathbb{Z}_p$, $j = 1, \ldots, t$, such that

$$f_j(T) = T^{\lambda_j} + p \times \text{(terms of lower degree)}$$

and finally a Λ-module map

$$\varphi: \mathcal{X}(F_\infty) \longrightarrow \Lambda^r \oplus \left(\bigoplus_{i=1}^{s} \Lambda/(p^{\mu_i})\right) \oplus \left(\bigoplus_{j=1}^{t} \Lambda/(f_j)\right)$$

with *finite kernel and cokernel*. We make the definitions

$$\mu_F \stackrel{\text{def}}{=} \textstyle\sum_{i=1}^{s} \mu_i,$$

$$\lambda_F \stackrel{\text{def}}{=} \sum_{j=1}^{t} \lambda_j;$$

we also define the polynomial

$$P_F \stackrel{\text{def}}{=} \prod_{j=1}^{t} f_j = T^{\lambda_F} + p \times (\text{terms of lower degree}),$$

and finally the *Iwasawa polynomial* of $\mathcal{X}(F_\infty)$,

$$Q_F(T) \stackrel{\text{def}}{=} p^{\mu_F} P_F(T).$$

Propositions 2.2 and 2.4 show that $r = 0$ and that $P_F(0) \neq 0$. (The first fact does not actually require Hypothesis 2.1 while the second is equivalent to it.) Note that μ_F, λ_F, P_F and Q_F depend only on F and the choice of γ. In particular, letting $V(F)$ denote the λ_F-dimensional $\bar{\mathbb{Q}}_p$-vector space $\mathcal{X}(F_\infty) \otimes_{\mathbb{Z}_p} \bar{\mathbb{Q}}_p$, it is not hard to show:

Lemma 2.5. *$P_F(T)$ is the characteristic polynomial of the endomorphism $1 - \gamma$ acting on $V(F)$.*

Next we have the important

Proposition 2.6. *$\mathcal{X}(F_\infty)$ has no non-zero, finite Λ-submodules.*

Proof See [12], Proposition 7. There $L/\mathbb{Q}$ is abelian and it is assumed that $[L(p) : L] = 2$. However, with trivial modifications the proof applies just as well to our more general situation. ◊ ◊

From this one deduces

Proposition 2.7. *In $I(\mathbb{Q}_p)$ we have*

$$|\mathcal{X}(F)| = (Q_F(0)) = (p^{\mu_F} P_F(0)). \qquad (2.2.1)$$

Proof The map φ has finite cokernel and is, by the previous proposition, injective. The result follows from Proposition 2.4 and a simple argument involving the Snake Lemma (see for example the proof of Lemma 6.3, p.538 of [6]). ◊ ◊

2.3 A factorizable function

The action of the group $\mathrm{Gal}(L/K)$ now enters the picture. We shall denote it G and assume that it is *abelian* (only because we have only defined the notion of factorizability for such groups). Because we are assuming Hypothesis 2.3, we can and shall identify G with $\mathrm{Gal}(L_\infty/K_\infty)$. Given a subgroup H of G we denote by $F = F(H)$ its fixed field and identify G/H with $\mathrm{Gal}(F_\infty/K_\infty)$. The latter, like the group Γ, acts on $\mathcal{X}(F_\infty)$ by extension and conjugation; moreover this action commutes with the Γ-action. Thus we may decompose $V(F)$ into eigenspaces for the action of G/H:

$$V(F) = \bigoplus_{\chi \in \widehat{G/H}} V(F)^\chi$$

and define $g_{F/K}(T, \chi)$ to be the characteristic polynomial of $1 - \gamma$ acting on $V(F)^\chi$. We then have:

Proposition 2.8. *For* $\chi \in \widehat{G/H}$ *and* $\omega \in \mathrm{Gal}(\bar{\mathbb{Q}}_p/\mathbb{Q}_p)$,

$$g_{F/K}(T, \chi^\omega) = g_{F/K}(T, \chi)^\omega$$

where ω *acts on the coefficients of* $g_{F/K}(T, \chi)$ *on the right-hand side.*

Furthermore

$$P_F(T) = \prod_{\chi \in \widehat{G/H}} g_{F/K}(T, \chi) \quad \text{in } \bar{\mathbb{Q}}_p[T]. \tag{2.3.2}$$

Proof The first statement is easily verified and the second is a consequence of Lemma 2.5. ◊ ◊

Let π_H denote the restriction map on Galois groups:

$$\pi_H : \mathcal{X}(L_\infty) \longrightarrow \mathcal{X}(F_\infty).$$

For each character χ of G/H, π_H clearly induces a map $V(L)^{\mathrm{Infl}\chi} \to V(F)^\chi$ and it is shown in [13], Proposition 1, that this map is an *isomorphism*. (See also Remark 4(a) below.) Since it commutes with the actions of Γ we obtain:

Proposition 2.9.

$$g_{F/K}(T, \chi) = g_{L/K}(T, \mathrm{Infl}\chi). \tag{2.3.3}$$

We can now prove the main result of this section.

Theorem 2.10. *Suppose that Hypothesis* 2.1 *holds. Then the function*

$$\begin{array}{ccc} \mathcal{S}(G) & \longrightarrow & I(\mathbb{Q}_p); \\ H & \longmapsto & (p^{-\mu_F}[M(F) : F_\infty]) \end{array}$$

where $F = L^H$, *is factorizable.*

Proof By Propositions 2.7, 2.8 and 2.9 we have

$$\begin{aligned}(p^{-\mu_F}[M(F):F_\infty]) &= (P_F(0))\\ &= \prod_{\psi\in\widehat{G/H}} (g_{F/K}(0,\psi))\\ &= \prod_{\substack{\chi\in\widehat{G}\\ \chi(H)=1}} (g_{L/K}(0,\chi))\end{aligned}$$

and

$$(g_{L/K}(0,\chi^\omega)) = (g_{L/K}(0,\chi)).$$

So Definition 1.3 is satisfied with $g(\chi) = (g_{L/K}(0,\chi))$. ◊ ◊

REMARK 4.

(a) The rôle played by Proposition 2.6 in the above is crucial. One can show that the map $\mathcal{X}(L_\infty)_H \to \mathcal{X}(F_\infty)$ induced by π_H has finite kernel and cokernel and hence that $\mathcal{X}(L_\infty)_H$ also has $Q_F(T)$ as its Iwasawa polynomial. Since $\mathcal{X}(L_\infty)_H$ *may have non-zero finite submodules,* however, the analogue of Proposition 2.7 fails. This distinction underlies the somewhat mysterious way in which the polynomials $g_{L/K}(T,\chi)$ for $\chi(H) = 1$ tell us about the size of $\mathcal{X}(F)$ (a module defined without reference to L) and not, for example, that of $\mathcal{X}(L)_H$.

(b) We have used equations (2.2.1), (2.3.2) and (2.3.3) in a similar way to equations (1.1.1), (1.1.2) and (1.1.3) in Section 2. One should point out that the evident similarities between these two sets of equations are far from coincidental. Indeed, the 'Main Conjecture of Iwasawa Theory' over K (now largely proven by Wiles) posits a close and specific connection between our $g_{F/K}(T,\chi)$ and certain p-adic L-functions. And these in turn are (roughly speaking) defined by p-adic interpolation at non-positive integers of the complex L-functions $L_{F(p)/K}(s,\psi)$ for various ψ. However, since we have not used (and will not use) the fact of this connection, we can regard the analogy as merely motivational.

We shall return to this result in the final section of the paper where it provides the basis for our main theorems on Galois module structure. The next section explains the way in which factorizability ties up with the structure of group lattices in general.

3 The defect function and factor equivalence

3.1 More definitions

We continue the notation of Section 1 and start by fixing a finite abelian group G and a prime p (not necessarily odd in this section). Φ denotes a $\mathbb{Q}_p$-conjugacy class of characters $\chi \in \widehat{G}$ and also, by abuse of notation, their pointwise sum. Thus the functions $\Phi: G \to \mathbb{Q}_p$ are precisely the characters of the irreducible $\mathbb{Q}_p$-representations of G. To each Φ there corresponds an idempotent e_Φ of the group-ring $\mathbb{Q}_pG$:

$$e_\Phi = \frac{1}{|G|}\sum_{g\in G}\Phi(g)g^{-1}$$

and $\mathbb{Q}_pG$ decomposes as a product of fields:

$$\mathbb{Q}_pG = \prod_{\Phi\in X_p(G)} e_\Phi\mathbb{Q}_pG.$$

Moreover, for each Φ, we have a commuting diagram:

$$\begin{array}{ccc} \mathbb{Q}_pG & \xrightarrow{\quad\chi\quad} & \mathbb{Q}_p(\Phi) \\ \big\downarrow & \nearrow_{\chi} & \\ e_\Phi\mathbb{Q}_pG & & \end{array}$$

Here, the vertical map is the projection induced by multiplication by e_Φ and χ is any element of Φ extended $\mathbb{Q}_p$-linearly. The slanting map is then an *isomorphism.*

An *order in* $\mathbb{Q}_pG$ is a $\mathbb{Z}_p$-lattice in $\mathbb{Q}_p$ which is also a subring of $\mathbb{Q}_pG$. Since G is abelian, all orders are contained in the unique maximal order $\mathfrak{M} = \mathfrak{M}(G)$ which is the integral closure of $\mathbb{Z}_p \in \mathbb{Q}_pG$. Of course, we have:

$$\mathfrak{M} = \prod_{\Phi\in X_p(G)} \mathcal{O}_\Phi \tag{3.1.1}$$

where $\mathcal{O}_\Phi$ is the ring of integers of $e_\Phi\mathbb{Q}_pG$. It is well known that the integers of $\mathbb{Q}_p(\Phi)$ are spanned over $\mathbb{Z}_p$ by the set of values of any $\chi \in \Phi$. Hence the linear extension of χ takes $\mathbb{Z}_pG$ onto these integers and we have:

$$\mathcal{O}_\Phi = e_\Phi\mathbb{Z}_pG$$

and thus:

$$\mathfrak{M} = \mathbb{Z}_pG \Longleftrightarrow e_\Phi \in \mathbb{Z}_pG\ \ \forall\Phi \Longleftrightarrow p \nmid |G|.$$

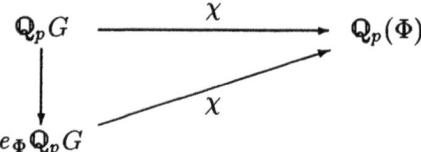

Now consider a $\mathbb{Z}_pG$-lattice $\mathfrak{A}$ contained in the $\mathbb{Q}_pG$-module $\mathfrak{A} \otimes \mathbb{Q}_p$. We write $\mathfrak{A}^{\mathfrak{M}}$ for the largest sublattice of $\mathfrak{A}$ on which $\mathfrak{M}$ acts. Since this contains $|G|\mathfrak{A}$ it is of finite index in $\mathfrak{A}$. One has, in fact:

$$\mathfrak{A}^{\mathfrak{M}} = \{a \in \mathfrak{A} : xa \in \mathfrak{A} \;\; \forall x \in \mathfrak{M}\} \cong \mathrm{Hom}_{\mathbb{Z}_pG}(\mathfrak{M}, \mathfrak{A}). \tag{3.1.2}$$

Now let $\mathfrak{B}$ be another $\mathbb{Z}_pG$-lattice.

Definition 3.1. *Suppose that $\mathfrak{B}$ is isogenous to $\mathfrak{A}$; then we define the* module defect *of $\mathfrak{A}$ and $\mathfrak{B}$ to be:*

$$\mathrm{j}^{\times}(\mathbb{Z}_pG, \mathfrak{A}, \mathfrak{B}) \stackrel{\mathrm{def}}{=} |\mathfrak{A} : A^{\mathfrak{M}(G)}|\, |\mathfrak{B} : B^{\mathfrak{M}(G)}|^{-1} \in I(\mathbb{Q}_p).$$

This satisfies certain obvious formal properties. For example, if $\mathfrak{A}$, $\mathfrak{B}$ and $\mathfrak{C}$ are mutually isogenous then we have:

$$\mathrm{j}^{\times}(\mathbb{Z}_pG, \mathfrak{A}, \mathfrak{B})\, \mathrm{j}^{\times}(\mathbb{Z}_pG, \mathfrak{B}, \mathfrak{C}) = \mathrm{j}^{\times}(\mathbb{Z}_pG, \mathfrak{A}, \mathfrak{C}). \tag{3.1.3}$$

Now for each subgroup H of G, $\mathfrak{A}^H$ and $\mathfrak{B}^H$ will be isogenous $\mathbb{Z}_p[G/H]$-lattices and one can form $(\mathfrak{A}^H)^{\mathfrak{M}(G/H)}$, etc. We make the definitions:

Definition 3.2. *The* defect function *of isogenous $\mathbb{Z}_pG$-lattices $\mathfrak{A}$ and $\mathfrak{B}$ is the function:*

$$\begin{array}{rcl} \mathrm{j}^{\times}(\mathfrak{A}, \mathfrak{B}) : \mathcal{S}(G) & \longrightarrow & I(\mathbb{Q}_p); \\ H & \longmapsto & \mathrm{j}^{\times}(\mathbb{Z}_p[G/H], \mathfrak{A}^H, \mathfrak{B}^H) \end{array}$$

and

Definition 3.3. *We shall say that $\mathbb{Z}_pG$-lattices $\mathfrak{A}$ and $\mathfrak{B}$ are* factor equivalent *(and write $\mathfrak{A} \wedge \mathfrak{B}$) if and only if they are isogenous and $\mathrm{j}^{\times}(\mathfrak{A}, \mathfrak{B})$ is a factorizable function.*

Equation (3.1.3) shows that this is indeed an equivalence relation between $\mathbb{Z}_pG$-lattices. It is intermediate in strength between isogeny and isomorphism. Note, however, that if $p \nmid |G|$ there is nothing of interest. For in this case $\mathbb{Z}_pG = \mathfrak{M}$ is a product of discrete valuation rings from which it follows that the relations of isogeny and isomorphism and hence also of factor equivalence coincide. (Indeed, $\mathrm{j}^{\times} \equiv 1$.)

Now suppose that $\mathfrak{A}$ and $\mathfrak{B}$ are isogenous. Then there exist $\mathbb{Z}_pG$-injections

$$i : \mathfrak{B} \longrightarrow \mathfrak{A}$$

and any such i induces for each $H \in \mathcal{S}(G)$ an injection:

$$i^H : \mathfrak{B}^H \longrightarrow \mathfrak{A}^H$$

with finite cokernel. We have:

Proposition 3.4. *The function*

$$\begin{array}{rcl} \mathcal{S}(G) & \longrightarrow & I(\mathbb{Q}_p); \\ H & \longmapsto & \mathrm{j}^{\times}(\mathfrak{A},\mathfrak{B})(H)^{-1}|\mathfrak{A}^H : i(\mathfrak{B}^H)| \end{array}$$

is factorizable.

Proof Briefly, the idea is to use the easily verified fact that

$$\mathrm{j}^{\times}(\mathfrak{A},\mathfrak{B})(H)^{-1}|\mathfrak{A}^H : i(\mathfrak{B}^H)| = |(\mathfrak{A}^H)^{\mathfrak{M}(G/H)} : i(\mathfrak{B}^H)^{\mathfrak{M}(G/H)}|.$$

By using equations (3.1.1) and (3.1.2) applied to the maximal order of $\mathbb{Z}_p[G/H]$, it is not hard to see that the right-hand side breaks up as a product of certain module indices over the set $\{\Phi \in X_p(G) : \Phi(H) = 1\}$. In order to get factorizability one then needs to show that these indices satisfy the norm condition 2 of the alternative form of Definition 1.3. For more details see [9], pages 411–412. ◊ ◊

The following proposition is a consequence of the 'Norm Theorem' which is proven in [9]:

Proposition 3.5. *As a function of H, $\mathrm{j}^{\times}(\mathfrak{A},\mathfrak{B})(H)$ is factorizable if and only if it is weakly factorizable.*

From the last two propositions one obtains:

Corollary 3.6. *The following are equivalent:*

1. $|\mathfrak{A}^H : i(\mathfrak{B}^H)|$ *is weakly factorizable as a function of H.*
2. $|\mathfrak{A}^H : i(\mathfrak{B}^H)|$ *is factorizable as a function of H.*
3. $\mathfrak{A} \wedge \mathfrak{B}$.

It follows that if condition 1 or 2 holds for one injection i then it will hold for any.

REMARK 5.

(a) Proposition 3.4 and Corollary 3.6 are extremely useful in calculations.

(b) The notions of the defect function and factor equivalence extend to arbitrary $\mathcal{R}G$-lattices with $\mathcal{R}$ as in Subsection 0.2. One can also define two isogenous $\mathcal{R}G$-lattices to be *weakly* factor equivalent if the corresponding defect function is weakly factorizable. (This is called 'factor equivalence' in [9].) Proposition 3.5 shows that in our situation

($\mathcal{R} = \mathbb{Z}_p$) such a definition would be superfluous. On the other hand, by means of the notion of a 'canonical' factorization one can define for each $\mathcal{R}$ an equivalence relation (canonical factor equivalence) which is considerably *stronger* than factor equivalence (see [3] or [20]). For the sake of simplicity we shall not pursue this here except to note that many of the results to be presented in Section 5 are capable of extension to the canonical situation as is shown in [20].

(c) Remark 1 shows that factor equivalence and isogeny coincide for cyclic G. More generally, one can show that this happens whenever the Sylow p-subgroup of G is cyclic. This indicates that factor equivalence is a much weaker relation than isomorphism in general.

EXAMPLE 5. It is shown in [9] that the maximal order $\mathfrak{M}(G)$ is itself factor equivalent to $\mathbb{Z}_pG$ *if and only if* the Sylow p-subgroup of G is cyclic.

EXAMPLE 6. Let L/K be a Galois extension of number fields with group G and let I be an ideal of $\mathcal{O}_K$. For each field F with $L \supseteq F \supseteq K$, write S_F for the set of prime ideals of $\mathcal{O}_F$ containing I and let

$$J_F \stackrel{\text{def}}{=} \bigcap_{\mathfrak{Q} \in S_F} \mathfrak{Q} = \prod_{\mathfrak{Q} \in S_F} \mathfrak{Q}.$$

Thus J_F is the *radical* of $I\mathcal{O}_F$, i.e.

$$J_F = \{x \in \mathcal{O}_F : x^n \in I\mathcal{O}_F \text{ for some } n \in \mathbb{N}\}.$$

Proposition 3.7. *For all primes p we have*

$$\mathcal{O}_{L,p} \wedge J_{L,p}.$$

(That is, $\mathcal{O}_L$ and J_L are factor equivalent as $\mathbb{Z}G$-lattices.)

Proof It is easily seen that J_L is a $\mathbb{Z}G$-sublattice of finite index in $\mathcal{O}_L$. Thus, by Corollary 3.6, it will suffice to show that the function

$$c_p : H \longmapsto |(\mathcal{O}_{L,p})^H : (J_{L,p})^H|$$

is factorizable.

Now it follows easily from the definition of J_L that $J_L^H = J_F$, where we write F for L^H, and thus

$$(\mathcal{O}_{L,p})^H/(J_{L,p})^H \quad \cong \quad (\mathcal{O}_F/J_F) \otimes \mathbb{Z}_p$$

$$\cong \bigoplus_{\mathfrak{Q} \in S_F} ((\mathcal{O}_F/\mathfrak{Q}) \otimes \mathbb{Z}_p)$$

by the 'Chinese Remainder Theorem' in $\mathcal{O}_F$. Hence we have

$$c_p(H) = \prod_{\substack{\mathfrak{Q} \in S_F \\ \mathfrak{Q}|p}} N\mathfrak{Q}.$$

Suppose that $\mathfrak{P}_K^{(1)}, \ldots, \mathfrak{P}_K^{(t)}$ are the prime ideals of $\mathcal{O}_K$ containing I and dividing p. Then, by adapting the notation of Example 4, the last equation may be written:

$$c_p(H) = \prod_{i=1}^{t} (N\mathfrak{P}_F^{(i)})^{g_F^{(i)}},$$

where for each i, $\mathfrak{P}_F^{(i)}$ is a chosen prime ideal of θ_F dividing $\mathfrak{P}_F^{(i)}$. Moreover that example shows that the right-hand side is a product of factorizable functions. Hence it is factorizable, as required. ◊ ◊

We note that $\mathcal{O}_L$ and J_L are not in general locally isomorphic as is shown by the following example of Ullom. Take $K = \mathbb{Q}$ and $L = \mathbb{Q}(i)$ where $i^2 = -1$. Then G is of order 2 and we take $I = 2\mathbb{Z}$ so that $J_L = (1+i)\mathcal{O}_L$. The reader may verify that J_L is $\mathbb{Z}G$-free of rank 1 on $1+i$. On the other hand, the prime 2 is wildly ramified in L, so $\mathcal{O}_L$ is not locally free at 2 by Theorem 0.1.

In the next subsection we present further (and rather deeper) instances of factorizability in an arithmetic setting.

3.2 Arithmetic applications

The two results stated here without proof apply the above theory respectively to the study of additive and multiplicative Galois module structure. They are slight reformulations of results in [10].

Let $L, K, G, n, S, \ldots$ be as in Subsection 0.2 and assume that G is *abelian*. Recall that $\mathcal{O}_{L,p}$ and $(\mathbb{Z}_pG)^n$ are isogenous. By contrast to Theorem 0.1, however, we also have:

Theorem 3.8. *The $\mathbb{Z}_pG$-lattices $\mathcal{O}_{L,p}$ and $(\mathbb{Z}_pG)^n$ are factor equivalent for all primes p.*

Proof See [10], Theorem 7 (Additive). The proof uses the theory of Galois Gauss sums. ◊ ◊

Turning to multiplicative Galois structure, we wish to compare the two $\mathbb{Z}_pG$-lattices $\mathfrak{U}_S(L)_p$ and $\mathfrak{H}_S(L)_p$ which are isogenous by Proposition 0.2. A finite place of K is said to *split* in L if there are precisely $|G|$ places of L lying above it. Each infinite place $v \in S_\infty(K)$ is either real with $|G|/2$ complex places above it in L, or else is real or complex with $|G|$ places above it of the same type as itself. By analogy with the finite places we shall say in the second case that v *splits* in L. Also, for $H \in \mathcal{S}(G)$ we shall denote by $F = F(H)$ the fixed field L^H and write $h_S(F)$ for $h_{S(F)}(F)$. We have:

Theorem 3.9. *Suppose that all the places in $S(K) \cup S_\infty(K)$ split in L and that $p \nmid (w_L, |G|)$. Then the expression*

$$\mathrm{j}^\times(\mathfrak{H}_S(L)_p, \mathfrak{U}_S(L)_p)(H) h_S(F) w(F)^{-1}$$

is weakly factorizable as a function of H.

Proof See [10], Theorem 7 (Multiplicative). ◊ ◊

In particular, using Proposition 3.5, we have:

Corollary 3.10. *Under the conditions of the theorem, $\mathfrak{H}_S(L)_p \wedge \mathfrak{U}_S(L)_p$ if and only if $h_S(F)w(F)^{-1} \in I(\mathbb{Q}_p)$ is weakly factorizable as a function of H.*

For observations on the factor $h_S(F)$, see subsection 4.2.

The proof of Theorem 3.9 in [10] uses, among other things, the complex L-function identities (1.1.1), (1.1.2) and (1.1.3) of Section 1. The factorizability result of Theorem 2.10 derived from their p-adic analogues will be applied in the next section to prove a result which is a 'hybrid' between Theorems 3.8 and 3.9. This will lead to more information on the Galois structure of $\mathfrak{U}_S(L)_p$ in certain cases.

4 Factor equivalence of S-units

4.1 Statement of the main results

Now let p be an *odd* prime number and L/K an extension of *totally real* number fields with abelian Galois group G. Let S denote a finite set of finite places of K. Using the results of previous sections, we shall obtain information on the Galois structure of $\mathfrak{U}_S(L)_p$ by comparing it to certain other $\mathbb{Z}_pG$-lattices which we now define.

For each field F with $L \supseteq F \supseteq K$, we shall denote by $\mathcal{O}(S,F)$ the direct sum $\mathcal{O}_F \oplus \mathbb{Z}^{S(F)}$ of $\mathbb{Z}[\mathrm{Gal}(F/K)]$-modules. Let $T_S(F)$ denote the map

$$\begin{array}{rcl} \mathcal{O}(S,F) & \longrightarrow & \mathbb{Z}; \\ (x,h) & \longmapsto & Tr_{F/\mathbb{Q}}(x) + \sum_{v\in S(F)} n_v h(v) \end{array}$$

where n_v is the local absolute degree of the place v. The kernel of $T_S(F)$ is a $\mathbb{Z}[\mathrm{Gal}(F/K)]$-lattice which we shall denote $\mathcal{O}(S,F)^0$. Also, we shall write $t_S(F)$ for the image of $T_S(F)$.

Next, let $\Sigma_S(F)$ denote the map of $\mathbb{Z}[\mathrm{Gal}(F/K)]$-modules

$$\begin{array}{rcl} \mathbb{Z}^{S_\infty(F)\cup S(F)} & \longrightarrow & \mathbb{Z}; \\ f & \longmapsto & \sum_{v\in S(F)\cup S_\infty(F)} n_v f(v) \end{array}$$

where $n_v = 1$, by definition, for an infinite place v. The kernel of $\Sigma_S(F)$ is a $\mathbb{Z}[\mathrm{Gal}(F/K)]$-lattice which we shall denote $\mathfrak{F}_S(F)$. Note that $\mathfrak{F}_S(F)$ and $\mathfrak{H}_S(F)$ coincide if $n_v = 1$ for all $v \in S(F)$ and that in general $\mathfrak{F}_S(L)$, $\mathcal{O}(S,L)^0$, $\mathfrak{H}_S(L)$ and $\mathfrak{U}_S(L)$ are all mutually isogenous $\mathbb{Z}G$-lattices.

To state the main theorem we recall the notation of previous sections and in addition we denote by $e_v(F/K)$ the ramification index of a place $v \in S(F)$ in the extension F/K.

Theorem 4.1. *Suppose that Hypothesis* 2.1 *holds and let* $F = F(H)$ *denote the fixed field* L^H *for* $H \in \mathcal{S}(G)$. *Then the expression*

$$\mathfrak{j}^\times(\mathcal{O}(S,L)^0_p, \mathfrak{U}_S(L)_p)(H)p^{-\mu_F}|\mathbb{Z}_p : t_S(F)_p|^{-1}h_S(F)\prod_{v\in S(F)} e_v(F/K)$$

is factorizable as a function of H.

We shall not prove this completely but will give an idea of the main steps in the proof in the final subsection.

The most direct application of Theorem 4.1 is perhaps the observation that the product of the last four terms in the given expression will be factorizable as a function of H if and only if the first term is. This in turn is, by definition, equivalent to the factor equivalence of the $\mathbb{Z}_pG$-lattices $\mathfrak{U}_S(L)_p$ and $\mathcal{O}(S,L)^0_p$. However, a rather neater factor equivalence result follows by combining Theorem 4.1 with Fröhlich's additive result (Theorem 3.8) as follows. From the latter one deduces:

Proposition 4.2. *The expression*

$$\mathrm{j}^{\times}(\mathfrak{F}_S(L)_p, \mathcal{O}(S,L)_p^0)(H)|\mathbb{Z}_p : t_S(F)_p|$$

is factorizable as a function of H.

(We omit the derivation which is not hard). Using equation (3.1.3) and Definition 3.2 we can now eliminate reference to the lattice $\mathcal{O}(S,L)_p^0$:

Theorem 4.3. *Suppose that Hypothesis* 2.1 *holds. Then the expression*

$$\mathrm{j}^{\times}(\mathfrak{F}_S(L)_p, \mathfrak{U}_S(L)_p)(H)p^{-\mu_F}h_S(F)\prod_{v\in S(F)} e_v(F/K)$$

is factorizable as a function of H.

Corollary 4.4. *Suppose that Hypothesis* 2.1 *holds. Then* $\mathfrak{F}_S(L)_p$ *is factor equivalent to* $\mathfrak{U}_S(L)_p$ *if and only if the expression*

$$p^{\mu_F}h_S(F)^{-1}\prod_{v\in S(F)} e_v(F/K)^{-1} \in I(\mathbb{Q}_p)$$

is factorizable as a function of H.

Remark 6. Proposition 3.5 shows that we could replace 'factorizable' by 'weakly factorizable' in Corollary 4.4.

4.2 Applications and comparison with Theorem 3.9

The comparison of the above results with Theorem 3.9 and its corollary is of interest not only for its own sake but also because it yields new information on the term involving the μ-invariant. This can then be 'ploughed back in' to improve the results.

In order to compare Theorems 3.9 and 4.3 we assume that the conditions of the latter are satisfied. Since L is then totally real and p is odd, the $w(F)$ term vanishes in Theorem 3.9 and also $S_\infty(K)$ is automatically split in L. Theorem 4.3 does not assume that $S(K)$ also splits. When this does hold however, the conditions of Theorem 3.9 are satisfied. Moreover, in this case $e_v(F/K) = 1$ for all F and v and one can show that $\mathfrak{F}_S(L)_p$ and $\mathfrak{H}_S(L)_p$ are *isomorphic* $\mathbb{Z}_pG$-lattices. Thus the two results agree if and only if the $p^{-\mu_F}$ term is weakly factorizable. But this term does not depend on S. So, taking for example $S = \varnothing$ (in which case $\mathfrak{F}_S(L) = \mathfrak{H}_S(L)$), we obtain:

Proposition 4.5. *Under the conditions of Theorem* 4.1, $p^{\mu_F} \in I(\mathbb{Q}_p)$ *is weakly factorizable as a function of* H.

It should be noted that μ_F is conjectured to be zero and that this is known to hold whenever F is abelian over $\mathbb{Q}$. (See [7] or [23], Section 7.5.)

Next we consider the problem of finding, for given L/K and various p, sets S such that Corollary 3.10 and Corollary 4.4 imply the factor equivalence of two lattices (that is, such that the specified functions from $\mathcal{S}(G)$ to $I(\mathbb{Q}_p)$ are (weakly) factorizable). We consider first Corollary 3.10 (thus L is no longer required to be totally real or p to be odd). Since factor equivalence is automatic if $p \nmid |G|$, we assume for now that p divides $|G|$ and attempt to ensure the weak factorizability at p of the two terms individually. Theorem 7 (Multiplicative) of [10] shows that the $w(F)$ term is weakly factorizable in $I(\mathbb{Q}_p)$ if p is odd. The most obvious way to ensure the weak factorizability of the $h_S(F)$ term at each prime dividing $|G|$ is to make S satisfy:

Condition 4.6. $(h_S(F), |G|) = 1$ *for all* F, $L \supseteq F \supseteq K$.

For example, we may take S to be *sufficiently large* by which we shall mean that $h_S(F) = 1$ for each F. (This is equivalent to saying that for each F, $S(F)$ contains primes whose images in $\mathrm{Cl}(F)$ generate.) Let us say that the extension L/K is *totally ramified* if and only if each subextension F'/F (with $L \supseteq F' \supseteq F \supseteq K$) unramified at every finite place, is trivial. Using Class Field Theory and the Čebotarev Density Theorem one can show:

There exists S split in K *and* satisfying Condition 4.6 if and only if L/K is totally ramified.

For non-totally ramified L/K it is not so clear how one might apply Corollary 3.10.

The situation with Corollary 4.4 is rather different. In the first place we need to assume that L is totally real, p is odd and that Hypothesis 2.1 is satisfied. However, since S is not required to split it is clearly possible to take it to be sufficiently large while avoiding the finitely many primes ramified in L/K. Then both the $h_S(F)$ term and the term in the $e_v(F/K)$ are trivial at any p. It follows from Theorem 4.3 that for such S, $\mathrm{j}^\times(\mathfrak{F}_S(L)_p, \mathfrak{U}_S(L)_p)(H) = p^{\mu_F}$ up to a factorizable function of H. By Proposition 4.5, p^{μ_F} and hence $\mathrm{j}^\times(\mathfrak{F}_S(L)_p, \mathfrak{U}_S(L)_p)$ are weakly factorizable as functions of H. Hence $\mathrm{j}^\times(\mathfrak{F}_S(L)_p, \mathfrak{U}_S(L)_p)$ is factorizable by Proposition 3.5. This implies that for such S, firstly $\mathfrak{F}_S(L)_p$ is factor equivalent to $\mathfrak{U}_S(L)_p$ and secondly p^{μ_F} is actually *factorizable*. However, μ_F is independent of S. Hence the following supersedes Proposition 4.5:

Proposition 4.7. *Under the conditions of Theorem* 4.1, $p^{-\mu_F} \in I(\mathbb{Q}_p)$ *is factorizable as a function of* H.

Corollary 4.8. *The term in* $p^{\pm\mu_F}$ *may be removed from the statements of Theorems* 4.1 *and* 4.3 *and Corollary* 4.4.

Finally, here is an example of an application:

Proposition 4.9. *Let L/K be any abelian extension of odd degree such that L is totally real and abelian over $\mathbb{Q}$. Let S be any sufficiently large, finite set of finite places of K which are unramified in L. Then we have*

$$\mathfrak{F}_S(L)_p \wedge \mathfrak{U}_S(L)_p \quad \textit{for all } p.$$

(We say that $\mathfrak{F}_S(L)$and $\mathfrak{U}_S(L)$are factor equivalent as $\mathbb{Z}G$-lattices.)

Proof Without loss of generality, $p \neq 2$ (since $[L : K]$ is odd). Since L is abelian over $\mathbb{Q}$, Hypothesis 2.1 is known to hold. (See [2] or [23], Section 5.5.) Now apply Corollary 4.4 noting that:

1. p^{μ_F} is factorizable in $I(\mathbb{Q}_p)$ by Proposition 4.7; in fact it is known to equal 1 since F is abelian over $\mathbb{Q}$.
2. $h_S(F) = 1$.

◊ ◊

4.3 Sketch proof of Theorem 4.1

For more details of the calculations in this section the reader should refer to the treatment of the 'canonical' case in [20].

The idea is to use Proposition 3.4 and for this we need to construct a $\mathbb{Z}_pG$-injection

$$\alpha : \mathfrak{U}_S(L)_p \longrightarrow \mathcal{O}(S, L)_p^0. \tag{4.3.1}$$

The construction utilizes the p-adic logarithm.

Some notations: For each field F with $L \supseteq F \supseteq K$, we shall denote $\mathrm{Gal}(L/F)$ by $H = H(F)$. For each prime ideal $\mathfrak{p}$ dividing p in $\mathcal{O}_F$, denote by $e_\mathfrak{p}$ and $\mathrm{ord}_\mathfrak{p}$ respectively the absolute ramification index of $\mathfrak{p}$ over p and the $\mathfrak{p}$-adic valuation. We shall denote by $\mathrm{ord}'_\mathfrak{p}$ the function

$$(1/e_\mathfrak{p})\mathrm{ord}_\mathfrak{p}\colon F^\times \longrightarrow (1/e_\mathfrak{p})\mathbb{Z} \subseteq \mathbb{Q}.$$

For $v \in S(F)$ we define ord'_v similarly. Also, for each F and $\mathfrak{p}$ we select a field embedding

$$\sigma_\mathfrak{p}\colon F \longrightarrow \bar{\mathbb{Q}}_p$$

such that the closure of the image of $\sigma_\mathfrak{p}$ is a completion of F at $\mathfrak{p}$, denoted $F_\mathfrak{p}$. (The $\sigma_\mathfrak{p}$ are chosen to be 'compatible' in the obvious sense as F and $\mathfrak{p}$ vary.) $Y_S(F)_p$ will denote the $\mathbb{Q}_p$ vector space

$$(F \otimes \mathbb{Q}_p) \oplus \mathbb{Q}_p{}^{S(F)}$$

on which $\mathrm{Gal}(F/K)$ acts. $\mathcal{O}(S,F)_p$ is then a $\mathbf{Z}_p[\mathrm{Gal}(F/K)]$-lattice in $Y_S(F)_p$ and $T_S(F)_p$ extends to a $\mathbf{Q}_pG$-linear map from $Y_S(F)_p$ to $\mathbf{Q}_p$. Note that the $\sigma_\mathfrak{p}$ induce an isomorphism

$$Y_S(F)_p \longrightarrow \left(\prod_{\mathfrak{p}|p} F_\mathfrak{p}\right) \oplus \mathbf{Q}_p{}^{S(F)}$$

which we shall regard as an identification.

Next, recall that the p-adic logarithm

$$\log_p : \bar{\mathbf{Q}}_p{}^{\times} \longrightarrow \bar{\mathbf{Q}}_p$$

is the unique homomorphism such that:

1. $\log_p(1+a) = a - \frac{1}{2}a^2 + \frac{1}{3}a^3 - \cdots$ if $|a|_p < 1$, and
2. $\log_p(p) = 0$.

It is continuous and $\mathrm{Gal}(\bar{\mathbf{Q}}_p/\mathbf{Q}_p)$-equivariant and hence takes $F_\mathfrak{p}^\times$ into a bounded subgroup of $F_\mathfrak{p}$ for each F and $\mathfrak{p}$. For $v \in S(F)$ let q denote the rational prime below v and define

$$\lambda_v = \begin{cases} 1 & \text{if } q = p \\ \log_p(q) & \text{otherwise.} \end{cases}$$

Now define a map

$$\beta_S(F): E_S(F) \longrightarrow Y_S(F)_p = \left(\prod_{\mathfrak{p}|p} F_\mathfrak{p}\right) \oplus \mathbf{Q}_p{}^{S(F)}$$

by

$$\beta_S(F)(x) = ((\mathrm{ord}'_\mathfrak{p}(x) + \log_p(\sigma_\mathfrak{p}(x)))_{\mathfrak{p}|p}, (-\lambda_v\, \mathrm{ord}'_v(x))_{v\in S(F)}).$$

One checks:

Proposition 4.10.

1. *$\beta_S(F)$ factors through $\mathfrak{U}_S(F)$ and extends $\mathbf{Z}_p$-linearly to a* $\mathrm{Gal}(F/K)$*-equivariant map which we shall denote $\beta_S(F)_p$:*

$$\beta_S(F)_p : \mathfrak{U}_S(F)_p \longrightarrow Y_S(F)_p$$

2. $\mathrm{Im}\beta_S(F)_p \subseteq \ker T_S(F)_p$
3. *Let F be the fixed field of $H \in \mathcal{S}(G)$. The inclusion $F \hookrightarrow L$ induces an injection $Y_S(F)_p \rightarrow Y_S(L)_p$ taking $\mathcal{O}(S,F)_p^0$ onto $(\mathcal{O}(S,L)_p^0)^H$. For each sufficiently large integer ξ,* $\mathrm{Im}(p^\xi\beta_S(L)_p)$ *is contained in $\mathcal{O}(S,L)_p^0$ and we have a commuting diagram:*

$$\begin{array}{ccc}
(\mathfrak{U}_S(L)_p)^H & \xrightarrow{(p^{\xi}\beta_S(L)_p)^H} & (\mathcal{O}(S,L)_p^0)^H \\
\Big\uparrow \wr\| & & \Big\uparrow \wr\| \\
\mathfrak{U}_S(F)_p & \xrightarrow{p^{\xi}\beta_S(F)_p} & \mathcal{O}(S,F)_p^0
\end{array}$$

Fix a ξ as in part 2 of Proposition 4.10 and for each F let

$$\alpha_S(F)_p : \mathfrak{U}_S(F)_p \longrightarrow \mathcal{O}(S,F)_p^0$$

denote $p^{\xi}\beta_S(F)_p$. We shall take α in (4.3.1) to be $\alpha_S(L)_p$. Then, by Proposition 4.10 and Proposition 3.4, the proof of Theorem 4.1 comes down to showing

$$\alpha_S(L)_p \text{ is injective} \tag{4.3.2}$$

and
In $I(\mathbb{Q}_p)$, up to a factorizable function of H:

$$\begin{aligned}&|\operatorname{coker} \alpha_S(F)_p| = \\ &\quad p^{\mu_F} |\mathbb{Z}_p : t_S(F)_p|\, h_S(F)^{-1} \textstyle\prod_{v \in S(F)} e_v(F/K)^{-1}\end{aligned} \tag{4.3.3}$$

The verification of these statements is achieved in three steps.

<u>STEP I: Reduce to the case $S = \varnothing$</u>

In this case we shall write $\mathcal{O}(F)_p^0$ for $\mathcal{O}(\varnothing, F)_p^0$, $t(F)_p$ for the ideal

$$t_{\varnothing}(F)_p = \operatorname{Tr}_{F/\mathbb{Q}}(\mathcal{O}_F) \otimes \mathbb{Z}_p \subseteq \mathbb{Z}_p$$

and so on. Now the assumption of Leopoldt's conjecture (Hypothesis 2.1) implies that $\alpha(F)_p$ is injective. Moreover, it is clearly induced by $\alpha_S(F)_p$ and the inclusions $\mathfrak{U}(F)_p \hookrightarrow \mathfrak{U}_S(F)_p$ and $\mathcal{O}(F)_p^0 \hookrightarrow \mathcal{O}(S,F)_p^0$. By examining the consequent map on the quotients we see that (4.3.2) holds and we obtain

In $I(\mathbb{Q}_p)$, up to a factorizable function of H:

$$\begin{aligned}&|\operatorname{coker} \alpha_S(F)_p| = \\ &\quad |\operatorname{coker}\ \alpha(F)_p|\ |t_S(F)_p : t(F)_p|^{-1} \tfrac{h(F)}{h_S(F)} \textstyle\prod_{v \in S(F)} e_v(F/K)^{-1}\end{aligned} \tag{4.3.4}$$

<u>STEP II: Calculate $|\operatorname{coker}\ \alpha(F)_p|$</u>

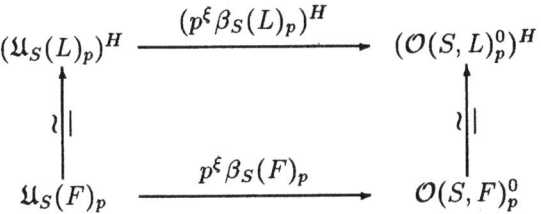

Since $p \neq 2$ and $W(F) = \{\pm 1\}$ we can identify $\mathfrak{U}(F)_p$ with $E(F)_p$ and this in turn identifies with $E^1(F)_p$. (Here, $E^1(F)$ denotes the group of those units of $\mathcal{O}_F$ which are congruent to 1 (that is, 'principal') modulo each prime ideal $\mathfrak{p}$ lying above p. The index $|E(F) : E^1(F)|$ is relatively prime to p). Also, write $U^1(F_p)$ for the product of the principal local units of F:

$$U^1(F_p) = \prod_{\mathfrak{p}|p} U^1(F_{\mathfrak{p}})$$

considered as a multiplicative $\mathbb{Z}_p$-module, and let $U^1(F_p)^0$ denote the kernel of the map

$$\mathcal{N}_{F/\mathbb{Q}}: U^1(F_p) \longrightarrow U^1(\mathbb{Q}_p) \subseteq \mathbb{Z}_p^\times$$

given by the product of the local norms. The embeddings $\sigma_{\mathfrak{p}}$ induce a map

$$j: E^1(F)_p \longrightarrow U^1(F_p)^0$$

whose image we denote $\overline{E^1(F)}$. To calculate the cokernel of $\alpha(F)_p$, we write the latter as the composite map

$$\begin{array}{rcl} \mathfrak{U}(F)_p = E^1(F)_p \xrightarrow{j} U^1(F_p)^0 & \xrightarrow{\Lambda} & \mathcal{O}(F)_p^0 \\ (x_{\mathfrak{p}})_{\mathfrak{p}|p} & \longmapsto & (p^{\xi} \log_p(x_{\mathfrak{p}}))_{\mathfrak{p}|p}. \end{array}$$

Using the properties of the p-adic logarithm, we obtain

In $I(\mathbb{Q}_p)$, up to a factorizable function of H:

$$|\operatorname{coker} \alpha(F)_p| = |U^1(F_p)^0 : \overline{E^1(F)}|\,|U^1(\mathbb{Q}_p) : \mathcal{N}_{F/\mathbb{Q}}(U^1(F_p))|^{-1}|\mathbb{Z}_p : t(F)_p| \quad (4.3.5)$$

<u>STEP III: Use the results of Section 2</u>

The proof of Theorem 4.1 will be complete once we have proven (4.3.3) above. Combining (4.3.4) with (4.3.5), it will suffice to show that the expression

$$p^{-\mu_F} |U^1(F_p)^0 : \overline{E^1(F)}|\ |U^1(\mathbb{Q}_p) : \mathcal{N}_{F/\mathbb{Q}}(U^1(F_p))|^{-1} h(F) \in I(\mathbb{Q}_p)$$

is factorizable as a function of H. Now, using Class Field Theory one can show that this expression is equal to

$$p^{-\mu_F} |\mathrm{Gal}(M(F)/F_\infty)|\ |\mathrm{Gal}(F \cap \mathbb{B}_\infty/\mathbb{Q})|^{-1}.$$

Under Hypothesis 2.3 the third factor here is constant as a function of H. (More generally, it can be shown to be factorizable.) By Theorem 2.10, the product of the first two factors is factorizable and this completes the proof of Theorem 4.1.

◇ ◇

References

1. R. Brauer, 'Beziehungen zwischen Klassenzahlen von Teilkörpern eines galoischen Körpers', Math. Nachr. **4** (1951), 158–174.

2. A. Brumer, 'On the units of algebraic number fields', Mathematika **14** (1967), 121–124.

3. D. Burns, 'Factorisability, group lattices and Galois module structure', Ph.D. Thesis, University of Cambridge (1990).

4. T. Chinburg, 'The analytic theory of multiplicative Galois structure', American Mathematical Society, Memoirs No. 395 (1989).

5. J. Coates, 'p-adic L-functions and Iwasawa's theory', in 'Algebraic Number Fields', Proceedings of the Durham Conference, ed. A. Fröhlich, Academic Press, London (1977).

6. J. Coates and S. Lichtenbaum , 'On l-adic zeta functions', Ann. Math. **98** (1973), 498–550.

7. B. Ferrero and L. Washington, 'The Iwasawa invariant μ_p vanishes for Abelian number fields', Ann. of Math. **109** (1979), 377–395.

8. A. Fröhlich, 'Galois module structure of algebraic integers', Springer-Verlag, Berlin–Heidelberg (1983).

9. A. Fröhlich, 'Module Defect and Factorisability', Illinois J. Math. **32** (1988), 407–421.

10. A. Fröhlich, 'L-values at zero and multiplicative Galois module structure (also Galois Gauss sums and additive Galois module structure)', J. reine angew. Math. **397** (1989), 42–99.

11. R. Gillard, 'Unités cyclotomiques, unités semi-locales et $\mathbf{Z}_\ell$-extensions', Ann. Inst. Fourier, Grenoble, **29, 1** (1979), 49–79.

12. R. Greenberg, 'On p-adic L-functions and cyclotomic fields II', Nagoya Math. J. **67** (1977), 139–158.

13. R. Greenberg, 'On p-adic Artin L-functions', Nagoya Math. J. **89** (1983), 77–87.

14. J.-F. Jaulent, 'Sur l'indépendance ℓ-adique de nombres algébriques', J. Number Theory **20** (1985), 149–158.

15. S. Lang, 'Algebraic Number Theory', Springer-Verlag, New York (1986).

16. S. Lang, 'Cyclotomic Fields', Springer-Verlag, New York (1978).

17. H.W. Leopoldt, 'Über die Hauptordnung der ganzen Elemente eines abelschen Zahlkörpers', J. reine angew. Math. **209** (1962), 54–71.

18. B. Mazur and A. Wiles, 'Class fields of Abelian extensions of $\mathbf{Q}$', Invent. Math. **76** (1984), 179–330.

19. A. M. Nelson, 'Monomial representations and Galois module structure', Ph.D. Thesis, University of London (1979).

20. D.R. Solomon, 'Canonical factorisations in multiplicative Galois structure', to appear in J. reine angew. Math.

21. J. Tate, 'Les conjectures de Stark sur les fonctions L d'Artin en $s = 0$', Prog. in Math. **47**, Birkhauser, (1984).

22. M. J. Taylor, 'On Fröhlich's conjecture for rings of integers of tame extensions', Invent. Math. **63** (1981), 41–79.

23. L. Washington, 'Introduction to Cyclotomic Fields', Springer-Verlag, New York (1982).

9.

WEAK FORMS OF AMENABILITY FOR SPLIT RANK 1 p-ADIC GROUPS

ALAIN VALETTE

Institut de Mathématiques,
Chantemerle 20,
CH-2007 Neuchâtel,
Switzerland.
`<mavalet@cnedcu51.bitnet>`

Introduction

Let $\mathbb{F}$ be a non-discrete locally compact field and let G be the group of $\mathbb{F}$-rational points of a simple algebraic group defined over $\mathbb{F}$. Let G have positive split rank, so that G is not compact in the locally compact topology coming from $\mathbb{F}$. For example, $G = \mathrm{SL}_n(\mathbb{F})$ when $n \geqslant 2$.

It is then a classical fact that G is not amenable. We recall the proof below. However, if $\mathbb{F}$ is a local field and G has split rank 1 (for example, $G = \mathrm{SL}_2(\mathbb{Q}_p)$, or $\mathrm{SL}_2(\mathbb{D})$, where $\mathbb{D}$ is a central division algebra over $\mathbb{Q}_p$), then weak forms of amenability do hold. The aim of this paper is to survey weak forms of amenability.

In Section 1, we quickly review some basic results about amenability: these involve various tools of harmonic analysis, and are needed in order to understand the generalizations presented in the sequel. In Section 2, we deal with the lack of Kazhdan's property (T), in Section 3 with K-theoretical amenability à la Cuntz, and in Section 4 with weak amenability à la Cowling–Haagerup. These three weak forms of amenability are satisfied by a split rank 1 simple algebraic group G over a local field, for a single geometrical reason: the group G has a proper action on some locally finite tree, the so-called 'Bruhat–Tits building of G'. This is to say that proofs only use this fact, and go over to other locally compact groups acting properly on locally finite trees. In this sense, this paper serves as an advertisement for harmonic analysis on trees. We also emphasize the fact that, because of the presence of the tree (a very simple geometrical object), the non-archimedean case is much easier to handle and to understand than the corresponding archimedean case: this will be illustrated in each section.

1 A quick survey of amenability

Throughout this paper, a *locally compact group* will be a Hausdorff, locally compact, second countable, topological group; $\mathbb{F}$ will denote a non-discrete locally compact field. A locally compact group G is *amenable* if any continuous affine action of G on a non-empty compact convex subset in a locally convex topological vector space admits a fixed point. For references about amenability, we recommend the books by Paterson [33], Pier [35], as well as Eymard's short and efficient paper [13].

Proposition 1.1. *Let G be a unimodular group. Assume G admits a closed, non-unimodular subgroup P such that G/P is compact. Then G is not amenable.*

Proof Because of the assumptions, there is no G-invariant measure on G/P. This means that the affine action of G on the compact convex set of probability measures on G/P has no fixed point. $\Diamond$ $\Diamond$

Corollary 1.2. *Let G be the group of $\mathbb{F}$-rational points of a simple algebraic group defined over $\mathbb{F}$. If G has positive split rank, then G is not amenable.*

Proof The group G is unimodular; on the other hand, since the split rank is positive, the group G contains parabolic subgroups. Such a subgroup P is non-unimodular, and G/P is compact. So Proposition 1.1 applies. $\Diamond$ $\Diamond$

The richness of the concept of amenability comes from the many equivalent definitions. We now review some of them.

A (unitary) *representation* of the locally compact group G is a homomorphism π from G to the unitary group $\mathcal{U}(\mathcal{H}_\pi)$ of some Hilbert space $\mathcal{H}_\pi$ such that the map $G \times \mathcal{H}_\pi \longrightarrow \mathcal{H}_\pi : (g,\xi) \longrightarrow \pi(g)\xi$ is continuous. The representation π is *irreducible* if it has no non-trivial closed invariant subspace in $\mathcal{H}_\pi$. Two representations are *equivalent* if there exists a unitary operator intertwining them. The set of classes of irreducible representations of G is the *dual* of G, denoted by $\hat{G}$. A distinguished element of the dual is the *trivial* one-dimensional representation, denoted by π_0.

A representation σ is *weakly contained* in a representation π if any positive definite coefficient function $g \longrightarrow < \sigma(g)\eta|\eta >$ associated with σ is the uniform limit over compact subsets of G of positive definite coefficient functions $g \longrightarrow < \pi(g)\xi|\xi >$ associated with π. In particular, π_0 is weakly contained in π if, for any $\varepsilon > 0$ and any compact subset K of G, there exists a unit vector ξ in $\mathcal{H}_\pi$ such that $||\pi(g)\xi - \xi|| < \varepsilon$ for any $g \in K$. Weak containment of representations can be used to define the *Fell topology* on $\hat{G}$; see [12], 18.15.

Fix a left Haar measure dg on G. It defines the Hilbert space $L^2(G)$, that carries the *left regular representation* of G, denoted by λ, and defined by $(\lambda(g)\xi)(x) = \xi(g^{-1}x)$ for any $\xi \in L^2(G)$ and any g, $x \in G$. The following theorem is a celebrated result of Hulanicki [23]; see also [48], 7.1.8 and [12], 18.3.6.

Theorem 1.3. *The following statements are equivalent:*

(i) G is amenable;

(ii) π_0 is weakly contained in λ;

(iii) Any representation π is weakly contained in λ.

The Banach space $L^1(G)$ becomes a Banach *-algebra when endowed with the convolution product and the involution

$$f^*(x) = \Delta(x^{-1})\overline{f(x^{-1})} \quad \text{for } x \in G \text{ and } f \in L^1(G),$$

where Δ denotes the modular homomorphism of G. Any representation π of G extends to a *-representation of $L^1(G)$, still denoted by π, via the formula:

$$\pi(f) = \int_G f(g)\pi(g)dg \qquad (f \in L^1(G)).$$

The *full C^*-algebra* of G, denoted by $C^*(G)$, is by definition the enveloping C^*-algebra of $L^1(G)$, i.e. the completion of $L^1(G)$ with respect to the norm

$$\begin{aligned} \|f\| &= \sup_{\pi \in \hat{G}} \|\pi(f)\| && (1.1) \\ &= \sup_{\pi \text{representation of} G} \|\pi(f)\| && (1.2) \end{aligned}$$

(for the equality of these norms, see [12], 2.7.1). On the other hand, one also has the *-representation λ of $L^1(G)$ (which is nothing but the action of $L^1(G)$ on $L^2(G)$ by left convolution!); the *reduced C^*-algebra* of G, denoted by $C_r^*(G)$, is the norm closure of $\lambda(L^1(G))$ in the space of all bounded linear operators on $L^2(G)$. The left regular representation induces a canonical map $\lambda : C^*(G) \longrightarrow C_r^*(G)$ that is onto.

More generally, given any C^*-algebra A endowed with a continuous action of G by *-automorphisms, one may construct a Banach *-algebra $L^1(G, A)$ (see [34], 7.6.1); the enveloping C^*-algebra of $L^1(G, A)$ is denoted by $A \rtimes G$, and is called the *full crossed product* of A by G ([34], 7.6.5). Taking any faithful representation ρ of A on some Hilbert space $CalH$, one obtains a faithful representation of $L^1(G, A)$ on the Hilbert space $L^2(G, \mathcal{H})$; the norm closure of $L^1(G, A)$ in the space of all bounded linear operators on $L^2(G, \mathcal{H})$ is the *reduced crossed product* of A by G; it does not depend on the choice of the faithful representation ρ, and it is denoted by $A \rtimes_r G$

([34], 7.7.4, 7.7.5). One also has the canonical map $\lambda_A: A \rtimes G \longrightarrow A \rtimes_r G$ which is onto. With the help of Theorem 1.3, it is not too difficult to prove the following (see [34], 7.3.9, 7.7.7):

Theorem 1.4. *The following statements are equivalent:*

(i) G is amenable;

(ii) The canonical map $\lambda: C^*(G) \longrightarrow C^*_r(G)$ is a *-isomorphism;

(iii) For any continuous action of G by *-automorphisms on a C^*-algebra A, the canonical map $\lambda_A: A \rtimes G \longrightarrow A \rtimes_r G$ is a *-isomorphism.

The *Fourier algebra* $A(G)$ of G is the space of coefficients of the left regular representation λ of G. Endowed with the pointwise product of functions and with the norm $\| \ \|_A$ coming from the identity

$$A(G) = L^2(G) * L^2(G)$$

(where $*$ denotes convolution), $A(G)$ becomes an abelian Banach algebra. An *approximate unit* in $A(G)$ is a net $(u_i)_{i \in I}$ in $A(G)$ such that for any $v \in A(G)$: $\lim_{i \in I} \|u_i v - v\|_A = 0$. Our final characterization of amenability is a famous result of Leptin [31]; see also [35], Theorem 10.4.

Theorem 1.5. *G is amenable if and only if $A(G)$ admits an approximate unit which is bounded in the norm $\| \ \|_A$.*

2 Lack of property (T).

A locally compact group G is said to have *Kazhdan's property* (T), or to be a *Kazhdan group*, if any representation of G that weakly contains the trivial representation π_0 actually contains it, i.e. admits non-zero G-fixed vectors.

It is an easy exercise that any compact group is Kazhdan. On the other hand, an amenable Kazhdan group G has to be compact (reason: combining the above definition with Theorem 1.3, one sees that $L^2(G)$ contains non-zero constant functions; this means that G is compact).

Apart from the obvious example of a compact group, proving that a given locally compact group G is Kazhdan is not an easy matter. All known constructions do use at some stage the following theorem of Kazhdan ([28]; see also [22], Theorem 2.8); we recall that a *lattice* Γ in G is a discrete subgroup such that G/Γ carries a finite G-invariant measure.

Theorem 2.1.

(i) Let G be the group of $\mathbf{F}$-rational points of a simple algebraic group of split rank $\geqslant 2$ defined over $\mathbf{F}$. Then G is Kazhdan.

(ii) If G is a Kazhdan group and Γ is a lattice in G, then Γ is Kazhdan.

In particular, the groups $\mathrm{SL}_n(\mathbf{R})$, $\mathrm{SL}_n(\mathbf{C})$, $\mathrm{SL}_n(\mathbf{Q}_p)$, $\mathrm{SL}_n(\mathbf{Z})$ are Kazhdan for $n \geqslant 3$. About the proof of Theorem 2.1(i), let us mention that it is a unified proof, working both for archimedean and non-archimedean fields, and that it involves no hard result in harmonic analysis. (Advertising sequence: the book [22] by de la Harpe and the author is maybe not the best but it *is* the most recent reference on property (T); the proof of Theorem 2.1(i) given there uses nothing but Bochner's theorem on positive definite functions on locally compact abelian groups, plus some standard material on the structure of algebraic groups.)

There is another equivalent definition of property (T), that allows one to deduce several important geometrical consequences. Before stating it, we need a definition: if X is a topological space, a continuous kernel $\Psi: X \times X \longrightarrow \mathbf{R}$ is *negative definite* if there exists a Hilbert space $\mathcal{H}$ and a continuous map $\beta: X \longrightarrow \mathcal{H}$ such that, for any $x, y \in X$:

$$\Psi(x, y) = \|\beta(x) - \beta(y)\|^2.$$

If G is a locally compact group, a continuous function $\psi : G \longrightarrow \mathbf{R}$ is *negative definite* if the kernel

$$G \times G \longrightarrow \mathbf{R};\ (g, h) \longrightarrow \psi(g^{-1}h)$$

is negative definite.

Theorem 2.2. *G is a Kazhdan group if and only if any continuous negative definite function on G is bounded.*

Like the AIDS virus, this result was first discovered by a French team (Delorme [11], Guichardet [17]), then rediscovered some time later by a US team (Akemann-Walter [1]). We first illustrate Theorem 2.2 on the example of trees. The next proposition was originally obtained in [25], lemma 2.3; it extends the fact, due to Haagerup [18], that the standard length function is negative definite on a non-abelian free group.

Proposition 2.3. *Let X be a tree with set of vertices Δ^0 and distance function d. Then d is a negative definite kernel on Δ^0.*

Proof Let Δ^1 be the set of edges of X. We choose a base-point $x_0 \in \Delta^0$, and get a map $\beta : \Delta^0 \longrightarrow \ell^2(\Delta^1)$ by defining $\beta(x)$ to be the characteristic function of the set of edges on the geodesic $[x_0, x]$. Using the special shape of triangles in a tree, it is easy to see that $\|\beta(x) - \beta(y)\|^2 = d(x, y)$ for all $x, y, \in \Delta^0$. ◊ ◊

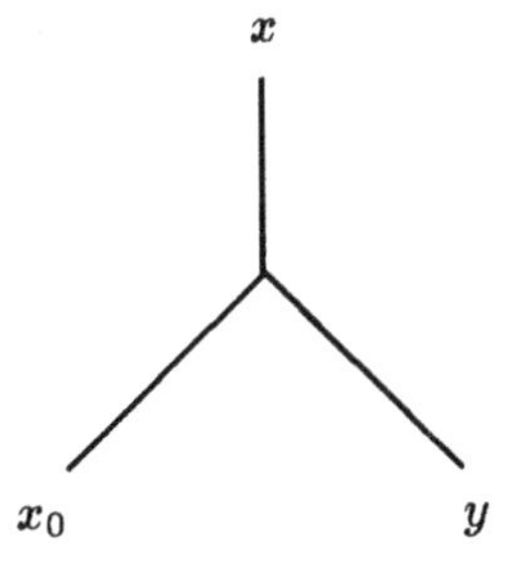

Corollary 2.4. *Let G be a Kazhdan group acting by automorphisms on a tree X. Then G stabilizes either some vertex or some edge of X.*

Proof Again, let X_0 be a base-point in Δ^0. The function $g \longmapsto d(x_0, gx_0)$ is negative definite on G, hence bounded. This means that the orbit of x_0 in Δ^0 is bounded. Let Y be the subtree spanned by this orbit: it is a G-invariant subtree of finite diameter N. Deleting all terminal vertices of Y (i.e. those vertices with only one neighbour in Y), we get a G-invariant subtree of diameter $N - 2$. Iterating this procedure, we eventually arrive at a G-invariant subtree of diameter 0 or 1, depending on the parity of N. In other words, G either has a fixed vertex or a fixed edge. ◊ ◊

The preceding argument can be found in [40], 1.4.3. Following Serre ([40], 1.6.6), we say that a group G has *property (FA)* if any orientation-preserving action of G on a tree has some fixed vertex. The above proof shows that property (T) implies property *(FA)*; the original proof of this can be found in [47].

Corollary 2.5. *Let G be a Kazhdan group; let H be the group of F-rational points of a split rank 1 simple algebraic group over a local field F. Any continuous homomorphism $G \longrightarrow H$ has relatively compact image; in particular, H is not Kazhdan.*

Before giving the proof, we notice that, if G is a lattice in a simple algebraic group of split rank $\geqslant 2$ over a non-discrete locally compact field, then Corollary 2.5 is a very special case of Margulis' Super-Rigidity Theorem (see [48], Chapter 5). In this sense, property (T) can be seen as a rigidity property.

Proof of Corollary 2.5 The Bruhat–Tits building, or building of affine type, associated with the group H is an oriented tree X on which H acts properly by orientation-preserving automorphisms (see [3]; actually the stabilizers in H of the vertices of X are precisely the maximal compact subgroups of H). The conclusion then follows from Corollary 2.4. $\diamondsuit\ \diamondsuit$

Let us now take a look at the archimedean counterpart of the above results. We now denote by d the hyperbolic distance either on the real hyperbolic n-space $H^n_{\mathbb{R}}$ or on the complex hyperbolic n-space $H^n_{\mathbb{C}}$. We quote without proof from a result of Faraut and Harzallah [14]:

Proposition 2.6. *d is a negative definite kernel either on $H^n_{\mathbb{R}}$ or on $H^n_{\mathbb{C}}$.*

From this one deduces the analogues of Corollaries 2.4 and 2.5.

Corollary 2.7. *Let G be a Kazhdan group endowed with a continuous action by isometries on $H^n_{\mathbb{R}}$ (resp. $H^n_{\mathbb{C}}$). Then G has a fixed point in $H^n_{\mathbb{R}}$ (resp. $H^n_{\mathbb{C}}$).*

Proof By Proposition 2.6, any orbit of G in $H^n_{\mathbb{R}}$ or $H^n_{\mathbb{C}}$ is bounded. Now, there are notions of convexity and barycentre in hyperbolic geometry: the closed convex hull of an orbit of G is a G-invariant compact convex subset; its barycentre is the desired fixed point. $\diamondsuit\ \diamondsuit$

Recall that the connected component of the isometry group of $H^n_{\mathbb{R}}$ (resp. $H^n_{\mathbb{C}}$) is locally isomorphic to the split rank 1 simple real Lie group $\mathrm{SO}_0(n,1)$ (resp. $\mathrm{SU}(n,1)$), and that the stabilizer of a point in hyperbolic space is a maximal compact subgroup in the group of isometries (see [21]). From this, we immediately get

Corollary 2.8. *Let G be a Kazhdan group. Any continuous homomorphism $G \longrightarrow \mathrm{SO}_0(n,1)$ or $G \longrightarrow \mathrm{SU}(n,1)$ has relatively compact image; in particular, $\mathrm{SO}_0(n,1)$ and $\mathrm{SU}(n,1)$ are not Kazhdan.*

Again, if G is a lattice in a simple algebraic group of split rank $\geqslant 2$ over $\mathbb{F}$, then Corollary 2.8 is a particular case of Margulis' super-rigidity. As an application of Corollaries 2.5 and 2.8, we mention a remarkable result of Zimmer [49].

Theorem 2.9. *Any countable Kazhdan subgroup in $\mathrm{SL}_2(\mathbb{C})$ is finite.*

Proof (sketch) We shall make use of the following facts: first, a countable Kazhdan group is finitely generated ([22], Theorem 1.11); second, any finitely generated torsion subgroup of a connected Lie group is finite ([39], corollary 6.14). So, if Γ is a countable Kazhdan subgroup of $\mathrm{SL}_2(\mathbb{C})$, all we have to show is that any element of Γ has finite order. Using the local isomorphism between $\mathrm{SL}_2(\mathbb{C})$ and $\mathrm{SO}_0(3,1)$, we may let Γ act on $H^3_{\mathbb{R}}$, and Corollary 2.8 tells us that Γ is relatively compact in $\mathrm{SL}_2(\mathbb{C})$. So we may assume that Γ is contained in $\mathrm{SU}(2)$. Then, to prove that a given $g \in \Gamma$ has finite order, it is enough to show that any eigenvalue λ of g is a root of unity. Let $\mathbb{K}$ be the subfield of $\mathbb{C}$ generated by λ and by all matrix entries of all elements of some finite generating set of Γ, so that $\Gamma \subseteq \mathrm{SL}_2(\mathbb{K})$. If λ is not a root of unity, we apply a lemma of Tits ([44], lemma 4.1) asserting that there exists a non-discrete locally compact field $\mathbb{F}$ and a non-zero homomorphism $\sigma : \mathbb{K} \longrightarrow \mathbb{F}$ such that $|\sigma(\lambda)| \neq 1$. Then the image of the homomorphism $\sigma : \Gamma \longrightarrow \mathrm{SL}_2(\mathbb{F})$ is not relatively compact, and this contradicts Corollary 2.5 if $\mathbb{F}$ is non-archimedean, and Corollary 2.8 if $\mathbb{F}$ is archimedean. ◊ ◊

Using Theorem 2.9, it is easy to check that any countable Kazhdan subgroup in $\mathrm{SO}(n)$ is finite, for $n = 2,3,4$. By way of contrast, for $n \geqslant 5$ there exist dense countable Kazhdan subgroups in $\mathrm{SO}(n)$: this was obtained independently by Margulis [32] and Sullivan [42]; see also [22], Proposition 3.7.

The list of split rank 1 simple Lie groups is given, up to local isomorphism, by $\mathrm{SO}_0(n,1)$, $\mathrm{SU}(n,1)$, $\mathrm{Sp}(n,1)$ and $F_{4(-20)}$ (see [21]): these are the isometry groups of hyperbolic n-spaces over the reals, the complex numbers, the Hamiltonian quaternions, and the Cayley octonions (with $n = 2$ in this last case). We have seen that $\mathrm{SO}_0(n,1)$, $\mathrm{SU}(n,1)$ are not Kazhdan; the really surprising fact is the following result, due to Kostant [29]:

Theorem 2.10. *The groups $\mathrm{Sp}(n,1)$ for $n \geqslant 2$ and $F_{4(-20)}$ are Kazhdan.*

In opposition with Theorem 2.1(i), this is truly a hard result: the few known proofs all involve deep results in harmonic analysis or representation theory (see [22], Chapter 9, for a discussion).

Dramatic applications of Theorem 2.10 have been given by Gromov, who proved that co-compact lattices in $\mathrm{Sp}(n,1)$ or $F_{4(-20)}$ admit uncountably many quotients that are infinite torsion groups with property (T) ([16], 5.5.E; this shows that Margulis' theorem on normal subgroups in higher rank lattices does not hold in rank 1); and by Skandalis, who used such lattices Γ to disprove the generalized Connes–Kasparov conjecture (this conjecture implies that $C_r^*(\Gamma)$ is K-theoretically equivalent to a nuclear C^*-algebra, and Skandalis showed this to be false in [41], Theorem 4.1).

Corollary 2.5 and Theorem 2.10 are in quite striking opposition, and illustrate the philosophy presented in the Introduction. Now, there is a result of Cowling that re-establishes some balance between the archimedean and non-archimedean cases. For a locally compact group G, property (T) is equivalent to the trivial representation π_0 being isolated in the dual $\hat{G}$ endowed with the Fell–Jacobson topology (see [22], Proposition 1.14). In [4], [5], Cowling deals with uniformly bounded representations; a continuous homomorphism π from G to the group of bounded invertible operators on a Hilbert space is a *uniformly bounded representation* if $\sup_{g\in G}\|\pi(g)\|$ is finite. Denote by $\hat{G}_{ub}$ the set of (classes of) irreducible uniformly bounded representations of G; this set $\hat{G}_{ub}$ can be given a topology extending the one on $\hat{G}$. Cowling proves:

Proposition 2.11. *Let G be a real simple Lie group with finite centre. The trivial representation π_0 is isolated in $\hat{G}_{ub}$ if and only if the split rank of G is not* 1.

3 K-amenability

For a locally compact group G, a *G-Fredholm module* is a triple (π^+, π^-, P), where π^+, π^- are (unitary) representations of G on Hilbert spaces $\mathcal{H}^+$, $\mathcal{H}^-$, and $P: \mathcal{H}^+ \longrightarrow \mathcal{H}^-$ is a bounded Fredholm operator intertwining π^+ and π^- modulo compact operators. G-Fredholm modules are the basic objects of Kasparov's equivariant KK-theory [27]; they form a semi-group under direct sum but, thanks to a suitable equivalence relation, Kasparov organized them into a unital commutative ring $R(G)$ which, for G compact, coincides with the usual representation ring. The unit 1_G of $R(G)$ is the class of the G-Fredholm module $(\pi_0, 0, 0)$ where π_0 is the one-dimensional trivial representation, and 0 the zero representation on the Hilbert space 0.

Kasparov's equivalence relation is built as follows: first, a G-Fredholm module (π^+, π^-, P) is *degenerate* if P is a unitary operator intertwining π^+ and π^-. Now, two G-Fredholm modules a, b are *equivalent* if there exists degenerate G-Fredholm modules x, y such that $a \oplus x$ is homotopic to $b \oplus y$. The homotopy relation involved here is somewhat subtle to describe. However, for our purpose it is enough to know the following: if $(\pi_t^+, \pi_t^-, P_t)_{t\in[0,1]}$ is a family of G-Fredholm modules, with all π_t^+ (resp. all π_t^-'s) acting on the same Hilbert space $\mathcal{H}^+$ (resp. $\mathcal{H}^-$), and if the maps

$$\begin{array}{rcl} G \times [0,1] & \longrightarrow & \mathcal{L}(\mathcal{H}^+ \oplus \mathcal{H}^-); \\ (g, t) & \longrightarrow & \pi_t^+(g) \oplus \pi_t^-(g) \end{array}$$

and

$$\begin{array}{rcl} [0,1] & \longrightarrow & \mathcal{L}(\mathcal{H}^+ \oplus \mathcal{H}^-); \\ t & \longrightarrow & \begin{pmatrix} 0 & P_t^* \\ P_t & 0 \end{pmatrix} \end{array}$$

are norm continuous, then the G-Fredholm modules (π_0^+, π_0^-, P_0) and $(\pi_1^+, \pi_1^-$ are certainly homotopic.

The locally compact group G is *K-amenable* if 1_G is equivalent to a G-Fredholm module (π^+, π^-, P) with π^+ and π^- weakly contained in the left regular representation λ. Of course, this definition is reminiscent of the characterization of amenability given in Theorem 1.3; clearly an amenable group is K-amenable. It can be said that the intuitive idea behind K-amenability is that the trivial representation π_0 is 'homotopic' to a difference of representations weakly contained in the left regular representation.

K-amenability was introduced by Cuntz [9], in order to explain conceptually the coincidence of the K-theory groups for the full and reduced C^*-algebras of free groups (see [8] and [38]). The following result was proved by Cuntz [9] (see also [25], corollary 3.6):

Proposition 3.1. *Let G be a locally compact group:*

(i) If G is K-amenable, then for any continuous action of G by *-automorphisms on a C^*-algebra A, the canonical map $\lambda_A : A \rtimes G \longrightarrow A \rtimes_r G$ induces isomorphisms in K-theory and in K-homology.

(ii) If G is discrete, the converse is true.

Proposition 3.1 is the K-theoretical analogue of Theorem 1.4; it is unknown whether (ii) in Proposition 3.1 holds without a discreteness assumption. With this proposition, we can show that K-amenability is a rather drastic strengthening of lack of property (T).

Proposition 3.2. *Let G be a non-compact group. If G is Kazhdan, then G is not K-amenable.*

Proof Because G is Kazhdan, π_0 is isolated in $\hat{G}$, as noticed at the end of Section 2. By [12], 3.2.3, this forces a direct sum decomposition $C^*(G) = \ker \pi_0 \oplus \mathbb{C}$, from which one sees that the idempotent e written $(0,1)$ in this decomposition generates a copy of $\mathbb{Z}$ in $K_0(C^*(G))$. Now,not being compact, G is not amenable (see the beginning of Section 2); by Theorem 1.3, π_0 is not weakly contained in λ, which implies $\lambda(e) = 0$. In particular, the map $\lambda_* : K_0(C^*(G)) \longrightarrow K_0(C_r^*(G))$ is not an isomorphism, which by Proposition 3.1 implies that G is not K-amenable. ◊ ◊

Up to now, the most general criterion for a group to be K-amenable has been given by the following theorem, due to Julg and the author ([25], Theorem 1.3):

Theorem 3.3. *Let G be a locally compact group acting on a tree, with amenable stabilizers of vertices. Then G is K-amenable.*

Let us mention that, for discrete groups, Pimsner [36] has obtained a very satisfactory extension of Theorem 3.3: if a discrete group G acts on a tree with K-amenable stabilizers of vertices, then G is K-amenable. Before giving some details about the proof of Theorem 3.3, let us mention one of its most interesting consequences ([25], Corollary 4.6):

Corollary 3.4. *Let H be the group of $\mathbb{F}$-rational points of a split rank 1 simple algebraic group over a local field. Then H is K-amenable.*

Proof of Corollary 3.4 As in Corollary 2.5, we use the action of H on its Bruhat–Tits building, which is a tree. The stabilizers of the vertices are compact, hence amenable. ◊ ◊

For the link between K-amenability of $\mathrm{SL}_2(\mathbf{Q}_p)$, potential theory on regular trees, and a classical problem in number theory, we refer to Haran's paper [20].

Proof of Theorem 3.3 The proof is in two steps:

1. *Construction of a G-Fredholm module γ_G.* Let X be a tree on which G acts with amenable stabilizers. We may assume that G preserves an orientation on X (otherwise, we replace X by its first barycentric subdivision). Let Δ^0 (resp. Δ^1) be the set of vertices (resp. edges) of X. We fix a base-point $x_0 \in \Delta^0$, and define a map $\beta: \Delta^0 - \{x_0\} \longrightarrow \Delta^1$ by sending a vertex x to the unique edge $\beta(x)$ through x that lies on the geodesic $[x_0, x]$. The map β is a bijection, and moreover, for $g \in G$, one has $\beta(gx) = g\beta(x)$ except if x lies on the finite subset $[g^{-1}x_0, x_0]$. Now, let λ_0 (resp. λ_1) be the natural representation of G on $\ell^2(\Delta^0)$ (resp. $\ell^2(\Delta^1)$), and let $(\delta_x)_{x\in\Delta^0}$ (resp. $(\delta_b)_{b\in\Delta^1}$) be the canonical basis of $\ell^2(\Delta^0)$ (resp. $\ell^2(\Delta^1)$). Define an operator $P: \ell^2(\Delta^0) \longrightarrow \ell^2(\Delta^1)$ by
$$P\delta_x = \begin{cases} 0 & \text{if } x = x_0 \\ \delta_{\beta(x)} & \text{if } x \neq x_0 \end{cases}$$
and notice that P extends linearly and continuously to a coisometry of Fredholm index 1; moreover, for any $g \in G$, the operator $P\lambda_0(g) - \lambda_1(g)P$ is finite rank, hence compact. This means that the triple $(\lambda_0, \lambda_1, P)$ is a G-Fredholm module γ_G. Because of the assumption that stabilizers are amenable, the representations λ_0, λ_1 are weakly contained in the left regular representation of G.

2. *Proof that $\lambda_G = 1_G$ in $R(G)$.* For this part of the proof, we follow Pimsner's approach [37], which is much simpler than the original one. For $x, y \in \Delta^0$ we denote by $\ell^2[x, y]$ the subspace of functions in $\ell^2(\Delta^0)$ with support in $[x, y]$ and by $\ell^2[x, y]^\perp$ the orthogonal subspace of $\ell^2[x, y]$. For $t \in [0, 1]$ and $(x, y) \in \Delta^1$, we introduce a unitary operator $c_t(x, y) \in \mathcal{U}(\ell^2(\Delta^0))$ by the formula
$$c_t(x, y) = \begin{pmatrix} \sqrt{1-t^2} & t \\ -t & \sqrt{1-t^2} \end{pmatrix} \oplus 1$$
in the decomposition
$$\ell^2(\Delta^0) = \mathbf{C}\delta_x \oplus \mathbf{C}\delta_y \oplus \ell^2[x, y]^1.$$
Caution: for this formula to make sense, one has to specify that the first basis vector of $\mathbf{C}\delta_x \oplus \mathbf{C}\delta_y$ is δ_x, while the second is δ_y. Using the elementary computation
$$\begin{pmatrix} 0 & 1 \\ 1 & 0 \end{pmatrix} \begin{pmatrix} \sqrt{1-t^2} & t \\ -t & \sqrt{1-t^2} \end{pmatrix} \begin{pmatrix} 0 & 1 \\ 1 & 0 \end{pmatrix} =$$

$$\begin{pmatrix} \sqrt{1-t^2} & t \\ -t & \sqrt{1-t^2} \end{pmatrix}^{-1}$$

one checks the relation $c_t(y,x) = c_t(x,y)^{-1}$. Now, if x, y are vertices at distance n, we denote by $x = y_0, y_1, y_2, \ldots, y_n = y$ the consecutive vertices on $[x, y]$, and we define

$$c_t(x,y) = c_t(y_0,y_1)c_t(y_1,y_2)\ldots c_t(y_{n-1},y_n).$$

In this way, we get a map $c_t : \Delta^0 \times \Delta^0 \longrightarrow \mathcal{U}(\ell^2(\Delta^0))$ which is a *cocycle* with respect to λ_0, in the following sense: for any $x, y, z \in \Delta^0$, $g \in G$ we have

$$\begin{aligned} c_t(x,x) &= 1, \\ c_t(x,y)c_t(y,z) &= c_t(x,z), \\ c_t(gx,gy) &= \lambda_0(g)c_t(x,y)\lambda_0(g^{-1}). \end{aligned}$$

Furthermore, in our case, c_t enjoys the important property that $c_t(x,y) - 1$ has finite rank. Now we define

$$\begin{aligned} \sigma_t : G &\longrightarrow \mathcal{U}(\ell^2(\Delta^0)) \\ g &\longmapsto c_t(x_0, gx_0)\lambda_0(g). \end{aligned}$$

Using the cocycle relations, it is easy to see that σ_t is a representation of G. Moreover, for any $g \in G$, the operator $\sigma_t(g) - \lambda_0(g)$ has finite rank, meaning that the triple (σ_t, λ_1, P) is a G-Fredholm module. It is clear from the definition that the map

$$\begin{aligned} G \times [0,1] &\longrightarrow \mathcal{L}(\ell^2(\Delta^0)) \\ (g,t) &\longmapsto \sigma_t(g) \end{aligned}$$

is norm continuous, hence the family $(\sigma_t, \lambda_1, P)_{t\in[0,1]}$ is a homotopy of G-Fredholm modules. To analyse the representation σ_1, we begin by using the orientation of the edges to define a map $\varepsilon : \Delta^1 \longrightarrow \{1, -1\}$ by

$$\varepsilon(b) = \begin{cases} 1 & \text{if } b \text{ points to } x_0, \\ -1 & \text{if not.} \end{cases}$$

Since G preserves the orientation of the edges, one has, for $g \in G$:

$$(*) \qquad \varepsilon(gb) = \varepsilon(b) \qquad \text{unless } b \subseteq [g^{-1}x_0, x_0].$$

Now define $\rho_1(g)\delta_b = \varepsilon(gb)\varepsilon(b)\lambda_1(g)\delta_b$ (for $b \in \Delta^1$, $g \in G$); one readily sees that ρ_1 is a representation of G on $\ell^2(\Delta^1)$. Now using the formulae of [45], lemma 2, one checks that $\sigma_1 = \pi_0 \oplus P^*\rho_1 P$, where the trivial representation π_0 takes place on $\mathbf{C}\delta_{x_0}$. This means that $\gamma_G = 1_G \oplus$

$(\rho_1, \lambda_1, 1)$, and it remains to show that $(\rho_1, \lambda_1, 1)$ is homotopic to a degenerate G-Fredholm module. To see that, consider, for $s \in [0,1]$, the diagonal operator

$$Q_s = \left(\operatorname{diag}\left[\exp \frac{i\pi s}{2}(1 - \varepsilon(b))\right]\right)_{b \in \Delta^1}.$$

Because of the almost invariance relation (∗), the operator Q_s intertwines ρ_1 and λ_1 modulo finite rank operators, i.e. (ρ_1, λ_1, Q_s) is a G-Fredholm module. Finally, one has $Q_0 = 1$, while $Q_1 = (\operatorname{diag} \varepsilon(b))_{b \in \Delta^1}$ intertwines ρ_1 and λ_1. This means that (ρ_1, λ_1, Q_1) is degenerate, and the proof is complete.

◊ ◊

We now discuss briefly the archimedean case. So, let G be a non-compact real simple Lie group with finite centre. Because of Proposition 3.2 and Theorems 2.1 and 2.10, we see that G is not K-amenable if G is not locally isomorphic either to $\mathrm{SO}_0(n,1)$ or to $\mathrm{SU}(n,1)$. The following theorem takes care of the remaining cases; it was obtained by Kasparov [26] for $\mathrm{SO}_0(n,1)$, and by Fox and Haskell [15] for $\mathrm{SU}(n,1)$:

Theorem 3.5. *Groups locally isomorphic either to* $\mathrm{SO}_0(n,1)$ *or* $\mathrm{SU}(n,1)$ *are K-amenable.*

In opposition with Theorem 3.3, whose proof only involves geometry of the underlying tree, the proof of Theorem 3.5 makes use of rather deep results in the representation theory of $\mathrm{SO}_0(n,1)$ and $\mathrm{SU}(n,1)$, namely the classification of complementary series.

It is worthwhile to notice that Kasparov in [26] proved much more that the K-amenability of $\mathrm{SO}_0(n,1)$: he showed that 'there is no Kasparov obstruction' for this group, meaning that one can reduce to a maximal compact subgroup in the computation of the K-theory of crossed products. Recently Julg and Kasparov [24] obtained the analogous statement for $\mathrm{SU}(n,1)$. So the precise result is as follows: let G be either $\mathrm{SO}_0(n,1)$ or $\mathrm{SU}(n,1)$; let K be a maximal compact subgroup of G; denote by p the dimension of the Riemannian symmetric space G/K. Then, for any continuous action of G by *-automorphisms on a C^*-algebra A, there is a canonical isomorphism in K-theory:

$$K_i(A \rtimes G) \cong K_{i+p}(A \rtimes K)$$

(and similarly for the K-homology groups).

To conclude this section, we mention a conjecture of Connes which, if true, would ensure that a group far enough from property (T) is automatically K-amenable. More precisely, assume that the locally compact group G admits a negative definite function φ such that $e^{-t\varphi}$ is integrable on G for t big enough; then G should be K-amenable.

4 Weak amenability

We have seen in Theorem 1.5 that amenability of the locally compact group G is equivalent to the existence of a bounded approximate unit in the Fourier algebra $A(G)$. In [18], Haagerup realized (in the case of non-abelian free groups) that, even for non-amenable groups, $A(G)$ might contain approximate units that are bounded in a weaker norm.

A function $f: G \longrightarrow \mathbf{C}$ is a *multiplier* of $A(G)$ if $fv \in A(G)$ for any $v \in A(G)$. With the operator norm, the space $M(G)$ of multipliers of $A(G)$ is a Banach algebra. A function $f \in M(G)$ is a *completely bounded multiplier* if and only if the following equivalent conditions are satisfied:

(a) For any locally compact group H, the function $f \otimes 1$ is contained in $M(G \times H)$;

(b) There exists a Hilbert space $\mathcal{H}$ and bounded continuous maps $P, Q \longrightarrow \mathcal{H}$ such that $f(g^{-1}h) = \langle P(g)|Q(h)\rangle$ for any $g, h \in G$.

The space of completely bounded multipliers is denoted by $M_0(G)$; the norms on $M_0(G)$ coming from definitions (a), (b) above do coincide, and $M_0(G)$ is a Banach algebra for this norm $\| \ \|_{M_0}$. Moreover, there are norm decreasing inclusions $A(G) \subseteq M_0(G) \subseteq M(G)$ (see [7], Section 0, for more on $M_0(G)$).

After Cowling and Haagerup [7], we say that G is *weakly amenable* if $A(G)$ admits approximate units that are bounded in the norm $\| \ \|_{M_0}$, i.e. if there exists $L \geqslant 1$ and a net $(u_i)_{i \in I}$ in $A(G)$ such that

$$\begin{cases} \lim_{i \in I} \|u_i v - v\|_A = 0 & \text{for any } v \in A(G) \\ \sup_{i \in I} \|u_i\|_{M_0} \leqslant L. \end{cases}$$

The *Cowling–Haagerup constant* Λ_G is the least upper bound of the set of L's when $(u_i)_{i \in I}$ runs among all nets having both properties above.

Theorem 4.1. *Let H be the group of $\mathbf{F}$-rational points of a split rank 1 simple algebraic group over a local field $\mathbf{F}$. Then H is weakly amenable, with Cowling–Haagerup constant $\Lambda_G = 1$.*

There are three different proofs of this result, all three based upon the geometry of trees. Szwarc's original proof [43] as well as the author's [46], follow from general results on groups acting properly on locally finite trees, while Bozejko and Picardello [2] obtain Theorem 4.1 as a consequence of the fact that an amalgamated product of amenable groups over a common compact open subgroup is weakly amenable with Cowling–Haagerup constant $\Lambda_G = 1$.

As in [46], we shall obtain Theorem 4.1 as a consequence of:

Proposition 4.2. *Let X be a tree, with set of vertices Δ^0; assume that the degrees of the vertices are bounded on Δ^0. Let G be a locally compact group acting properly on X. Then G is weakly amenable, with Cowling–Haagerup constant $\Lambda_G = 1$.*

We mention that in Szwarc's proof ([43], Theorem 6), the assumption of boundedness of the degrees of the vertices is superfluous.

Proof (Sketch) Denote by N the maximal degree of vertices in Δ^0. Let z be a complex number with $Re z > 0$. For an edge (x, y) in X we define an invertible operator $c_z(x, y)$ on $\ell^2(\Delta^0)$ by

$$c_z(x,y) = \begin{pmatrix} \sqrt{1-e^{-2z}} & e^z \\ -e^z & \sqrt{1-e^{-2z}} \end{pmatrix} \oplus 1$$

in the decomposition $\ell^2(\Delta^0) = \mathbf{C}\delta_x \oplus \mathbf{C}\delta_y \oplus \ell^2[x,y]^\perp$. As in the proof of Theorem 3.3, we extend c_z to a cocycle on X, the only difference being that c_z takes values in the general linear group of $\ell^2(\Delta^0)$ instead of the unitary group. Fixing a base-point $x_0 \in \Delta^0$, we define a non-unitary representation τ_z of G by $\tau_z(g) = c_z(x_0, gx_0)\lambda_0(g)$. By proposition 3 and lemma 2 of [45], the representation τ_z is uniformly bounded, it is unitary for z real, and the coefficient φ_z of τ_z on δ_{x_0} is given by:

$$\varphi_z(g) = \langle \tau_z(g)\delta_{x_0} | \delta_{x_0} \rangle = e^{-zd(x_0,gx_0)} \quad (\text{for } g \in G).$$

For $z, \delta > 0$, define $\varphi_{z,\delta}: G \longrightarrow \mathbf{C}$ by:

$$\varphi_{z,\delta}(g) = \frac{1}{\delta}\int_{-\infty}^{\infty} \exp(-\pi t^2\delta^{-2})\varphi_{z(1+it)}(g)\, dt.$$

We claim that $\varphi_{\delta,z}$ belongs to $A(G)$. To see it, we follow an idea of Cowling [6]. First, it follows from proposition 5 of [45] that, for $\mathrm{Re}\, z > \frac{1}{2}\log(N-1)$, the function φ_z is a coefficient of the natural representation λ_0 of G on $\ell^2(\Delta^0)$. Because the action of G on X is proper, any coefficient of λ_0 is in $A(G)$; hence $\varphi_z \in A(G)$ for $\mathrm{Re}\, z > \frac{1}{2}\log(N-1)$. Then, assume z is real, and fixed: let $w > 0$ be such that $wz > \frac{1}{2}\log(N-1)$. For $\delta > 0$, consider the entire function

$$\begin{aligned} f_\delta: \mathbf{C} &\longrightarrow \mathbf{C}; \\ s &\longrightarrow \frac{1}{\delta}\exp[\pi(s-1)^2\delta^{-2}] \end{aligned}$$

and the rectangular contour C_T in $\mathbf{C}$ with vertices $1 \pm iT$, $w \pm iT$ (with $T > 0$). By Cauchy's Theorem:

$$\oint_{C_T} f_\delta(s)\varphi_{zs}(g)\, ds = 0.$$

Letting T go to infinity, we get:

$$\varphi_{z,\delta}(g) = \frac{1}{\delta} \int_{-\infty}^{\infty} \exp[\pi(w-1+it)^2\delta^{-2}]\varphi_{z(w+it)}(g)\, dt$$

The right-hand term defines a function in $A(G)$; indeed, by the preceding remarks, the function $t \longrightarrow \varphi_{z(w+it)}$ is a periodic continuous function from $\mathbb{R}$ to $A(G)$. Hence the integral converges in $A(G)$, which establishes the claim.

It was proved in [7], Proposition 1.1, that a locally compact group G is weakly amenable if and only if there exists $L \geqslant 1$ and a net $(u_i)_{i\in I}$ in $A(G)$, converging to 1 uniformly on compact subsets of G, and such that $(u_i)_{i\in I}$ is bounded by L in the norm $\|\ \|_{M_0}$; moreover Λ_G is the least upper bound of these L, taken over all nets $(u_i)_{i\in I}$. Thus, to conclude, we let δ go to zero and notice that $\varphi_{z,\delta}$ converges to φ_z uniformly on compact subsets of G; then, letting z go to zero, φ_z converges to 1 uniformly on compact subsets of G. This shows the weak amenability of G. Finally, since φ_z is positive definite (recall that z is now real!), we have $\|\varphi_z\|_{M_0} = 1$, hence $\Lambda_G = 1$. ◊ ◊

Concerning weak amenability for non-compact simple real Lie groups, one has the following:

Theorem 4.3.

(i) *Let G be a split rank 1 simple real Lie group. Then G is weakly amenable, with the following Cowling–Haagerup constants:*
$\Lambda_G = 1$ *if G is locally isomorphic to* $\mathrm{SO}_0(n,1)$ *or* $\mathrm{SU}(n,1)$;
$\Lambda_G = 2n-1$ *if G is locally isomorphic to* $\mathrm{Sp}(n,1)$;
$\Lambda_G = 21$ *if G is the exceptional group* $F_{4(-20)}$.

(ii) *A simple real Lie group with split rank $\geqslant 2$ and finite centre is not weakly amenable.*

Many hands contributed to Theorem 4.3: the case of $\mathrm{SO}_0(n,1)$ was treated by De Cannière and Haagerup [10], while Cowling [6] dealt with groups with finite centre locally isomorphic to $\mathrm{SU}(n,1)$. The difficult computation for $\mathrm{Sp}(n,1)$ and $F_{4(-20)}$ was done by Cowling and Haagerup [7]. Recently, Lemvig Hansen [30] disposed of the remaining split rank 1 group, namely the universal covering group of $\mathrm{SU}(n,1)$. Once more, in opposition with the p-adic case, the proofs use some heavy machinery from representation theory, $\mathrm{Sp}(n,1)$ and $F_{4(-20)}$ being especially tough.

Part (ii) of Theorem 4.3 is an unpublished result of Haagerup [19]. Apparently, nobody attacked the case of p-adic groups with split rank $\geqslant 2$, but we are ready to bet that the conclusion will be the same as in the real case. We refer to [7], Section 6, for the important applications of Theorem 4.3 to group von Neumann algebras and II_1-factors.

Reaching the end of this paper, it is interesting to compare the results we have given on lack of property (T), K-amenability and weak amenability. Comparing Proposition 2.11 with Theorem 4.3, it would be tempting to 'explain' weak amenability of G by the fact that π_0 is not isolated in $\widehat{G}_{ub}$; however, this explanation does not transpire from the proofs of weak amenability. Similarly, one notices that the class of K-amenable groups seems to coincide with the class of weak amenable groups with Cowling–Haagerup constant $\Lambda_G = 1$. It is a challenging question to try to work out the precise relation between lack of property (T), K-amenability and weak amenability.

References

1. C.A. Akemann and M.E. Walter, 'Unbounded negative definite functions', Can. J. Math. **33** (1981), 862–871.

2. M. Bozejko and M. Picardello, 'Weakly amenable groups and amalgamated products', preprint (1989).

3. F. Bruhat and J. Tits, 'Groupes réductifs sur un corps local (données radicielles valuées)', Publ. Math. IHES **41** (1972), 5–252.

4. M. Cowling, 'Sur les coefficients des représentations des groupes de Lie simples', in 'Analyse harmonique sur les groupes de Lie II', Springer Lecture Notes in Math. **739** (1979), 132–178.

5. M. Cowling, 'Unitary and uniformly bounded representations of some simple Lie groups', in 'Harmonic Analysis and Group Representations, CIME 1980' (Liguori Napoli, 1982) 49–128.

6. M. Cowling, 'Harmonic analysis on some nilpotent groups', in 'Topics in Modern Harmonic Analysis', Vol. I, 81–123, Instituto Nazionale di Alta Matematica, Roma (1983).

7. M. Cowling and U. Haagerup, 'Completely bounded multipliers of the Fourier algebra of a simple Lie group of real rank one', Invent. Math. **96** (1989), 507–549.

8. J. Cuntz, 'The K-groups for free products of C*-algebras', Proc. Symp. Pure Math. **38** (1982), part 1, 81–83.

9. J. Cuntz, 'K-theoretic amenability for discrete groups', J. reine angew. Math. **344** (1983), 180–195.

10. J. De Cannière and U. Haagerup, 'Multipliers of the Fourier algebra of some simple Lie groups and their discrete subgroups', Amer. J. Math. **107** (1984), 455–500.

11. P. Delorme, '1-cohomologie des représentations unitaires des groupes de Lie semi-simple et résolubles; produits tensoriels continus et représentations', Bull. Soc. Math. France **105** (1977), 281–336.

12. J. Dixmier, 'Les C^*-algèbres et leurs représentations', 2ème édition, Gauthier-Villars, (1969).

13. P. Eymard, 'Introduction à la théorie des groupes moyennables', in 'Analyse harmonique sur les groupes de Lie', Springer Lecture Notes in Math. **497** (1975), 89–107.

14. J. Faraut and K. Harzallah, 'Distances hilbertiennes invariantes sur un éspace homogène', Ann. Inst. Fourier **24-3** (1974), 171–217.

15. J. Fox and P. Haskell, 'K-amenability for $\mathrm{SU}(n,1)$', to appear in J. Funct. Anal.

16. M. Gromov, 'Hyperbolic groups', in 'Essays in group theory', (ed. by S.M. Gersten), Springer (1987), 75–263.

17. A. Guichardet, 'Étude de la 1-cohomologie et de la topologie du dual pour les groupes de Lie à radical abelien', Math. Ann. **228** (1977), 215–232.

18. U. Haagerup, 'An example of a non-nuclear C^*-algebra which has the metric approximation property', Invent. Math. **50** (1979), 279–293.

19. U. Haagerup, 'Group C^*-algebras without the completely bounded approximation property', Unpublished manuscript (May 1986).

20. S. Haran, 'Index theory, potential theory, and the Riemann hypothesis', to appear in 'Proc. LMS Symp. on L-functions and Arithmetic', Durham, 1989.

21. S. Helgason, 'Differential geometry, Lie groups and symmetric spaces', Academic Press, 1978.

22. P. de la Harpe and A. Valette, 'La propriété (T) de Kazhdan pour les groupes localement compacts', Astérisque **175** (1989).

23. A. Hulanicki, 'Means and Følner conditions on locally compact groups', Studia Math. **27** (1966), 87–104.

24. P. Julg and G. G. Kasparov, 'L'anneau $KK_G(\mathbb{C},\mathbb{C})$ pour $G = \mathrm{SU}(n,1)$', to appear in CR Acad. Sci. Paris.

25. P. Julg and A. Valette, 'K-theoretic amenability for $\mathrm{SL}_2(\mathbb{Q}_p)$ and the action on the associated tree', J. Funct. Anal. **58** (1984), 194–215.

26. G. G. Kasparov, 'Lorentz groups: K-theory of unitary representations and crossed products', Dokl. Akad. Nauk SSSR **275** (1984), 541–545.

27. G. G. Kasparov, 'Equivariant KK-theory and the Novikov conjecture', Invent. Math. **91** (1988), 147–201.

28. D. Kazhdan, 'Connection of the dual space of a group with the structure of its closed subgroups', Funct. Anal. Appl. **1** (1967) , 63–65.

29. B. Kostant, 'On the existence and irreducibility of certain series of representations', Bull, Amer. Math. Soc. **75** (1969), 627–642.

30. M. Lemvig Hansen, 'Weak amenability of the universal covering group of SU$(n,1)$', Math. Ann. **288** (1990), 445–72.

31. H. Leptin, 'Sur l'algèbre de Fourier d'un groupe localement compact', CR Acad. Sci. Paris, Sér A, **266** (1968), 1180–1182.

32. G.A. Margulis, 'Some remarks on invariant means', Mh. Math. **90** (1980), 233–235.

33. A. L. Paterson, 'Amenability', Amer. Math. Soc. Surv. Monogr. **29** (1988).

34. G. K. Pedersen, 'C^*-algebras and their automorphism groups', Academic Press, (1979).

35. J.-P. Pier, 'Amenable locally compact groups', Wiley-Interscience, (1984).

36. M. Pimsner, 'KK-groups of crossed products by groups acting on trees', Invent. Math. **86** (1986), 603–634.

37. M. Pimsner, 'Cocycles on trees', J. Operator Theory **17** (1987), 121–128.

38. M. Pimsner and D. Voiculescu, 'K-groups of reduced crossed products by free groups', J. Operator Theory **8** (1982), 131–156.

39. M. S. Raghunathan, 'Discrete subgroups of Lie groups', Springer, (1979).

40. J.-P. Serre, 'Arbres, amalgames, SL_2', Astérisque **46** (1977).

41. G. Skandalis, 'Une notion de nucléarité en K-théorie', K-theory **1** (1988), 549–573.

42. D. Sullivan, 'For $n > 3$ there is only one finitely additive rotationally invariant measure on the n-sphere defined on all Lebesgue measurable subsets', Bull. Amer. Math. Soc. **4** (1981), 121–123.

43. R. Szwarc, 'Groups acting on trees and approximation properties of the Fourier algebra', J. Funct. Anal. **95** (1991), 320–343.

44. J. Tits, 'Free subgroups in linear groups', J. Algebra **20** (1972), 250–270.

45. A. Valette, 'Cocycles d'arbres et représentations uniformément bornées', CR Acad. Sci. Paris, **310** Sér. I (1990), 703–708.

46. A. Valette, 'Les représentations uniformément bornées associées à un arbre réel', Bull. Soc. Math. Belgique, Sér. A, **XLII** (1990), 747–760.

47. Y. Watatani, 'Property (T) of Kazhdan implies property *(FA)* of Serre', Math. Japon. **27** (1981), 97–103.

48. R. J. Zimmer, 'Ergodic theory and semisimple groups', Birkhäuser, (1984).

49. R. J. Zimmer, 'Kazhdan groups acting on compact manifolds', Invent. Math. **75** (1984), 425–436.

10.

p-ADIC FOURIER SERIES

C. F. WOODCOCK

Institute of Mathematics and Statistics,
The University,
Canterbury,
Kent CT2 7NF,
England.
<cfw@uk.ac.ukc>

Introduction

Throughout $\mathbb{Z}_p$ and $\mathbb{Q}_p$ will respectively denote the ring of p-adic integers and the field of p-adic numbers (for p prime). Thus $\mathbb{Z}_p$ is compact and is the closure of $\mathbb{Z}$ (indeed of $\mathbb{N}$) in $\mathbb{Q}_p$. We denote by $\mathbb{C}_p$ the completion of the algebraic closure of $\mathbb{Q}_p$ with respect to the p-adic metric. Let v_p denote the p-adic valuation of $\mathbb{C}_p$ normalized so that $v_p(p) = 1$.

Then $\mathbb{C}_p$ is the p-adic analogue of the field of complex numbers $\mathbb{C}$ and is the 'smallest' complete and algebraically closed field containing $\mathbb{Q}_p$ with the p-adic metric. However, unlike $\mathbb{C}$, it is not locally compact; in fact the locally compact subfields of $\mathbb{C}_p$ are precisely the finite extensions K of $\mathbb{Q}_p$ in $\mathbb{C}_p$ and these have dense union in $\mathbb{C}_p$. If $[K : \mathbb{Q}_p] = n$ and $x \in K$ has norm $N(x) \in \mathbb{Q}_p$, the norm being taken from K to $\mathbb{Q}_p$, then $v_p(x) = v_p(N(x))/n$ (the p-adic analogue of $|z|^2 = z\overline{z}$ for $z \in \mathbb{C}$).

We now introduce two subgroups $\mathbb{U}_p$, $\mathbb{T}_p$ of the multiplicative group $(\mathbb{C}_p^\times, \times)$ of $\mathbb{C}_p$ which will play an important role in the sequel:

Put $\mathbb{U}_p = \{\beta \in \mathbb{C}_p | v_p(\beta - 1) > 0\}$ and $\mathbb{V}_p = \mathbb{C}_p \backslash \mathbb{U}_p$. Then $(\mathbb{U}_p, \times)$ is a subgroup of $(\mathbb{C}_p^\times, \times)$, the group of so-called principal units (clearly $v_p(\beta) = 0$ for all $\beta \in \mathbb{U}_p$). Further $(\mathbb{U}_p, \times)$ admits a natural ('power') $\mathbb{Z}_p$-module structure which we will write exponentially, viz. β^z for $\beta \in \mathbb{U}_p$ and $z \in \mathbb{Z}_p$. Indeed if $\{n_k\}$ is a sequence of integers (or natural numbers) with $n_k \longrightarrow z$ in $\mathbb{Z}_p$ then in $\mathbb{U}_p$ we have

$$\beta^z = \lim_{k \longrightarrow \infty} \beta^{n_k}.$$

Put $\mathbf{T}_p = \{\omega \in \mathbf{C}_p | \omega^{p^n} = 1 \text{ for some } n \geqslant 0\}$. Then $\mathbf{T}_p \subseteq \mathbf{U}_p$ is the union of cyclic (multiplicative) groups C_{p^n} of order p^n $(n \geqslant 0)$ and has the discrete topology. In fact if $\omega \in \mathbf{T}_p$ has order p^n for some $n \geqslant 1$, then $v_p(\omega - 1) = 1/p^{n-1}(p-1)$. Then $(\mathbf{T}_p, \times)$ is the subgroup of $(\mathbf{U}_p, \times)$ consisting of all roots of unity in $\mathbf{U}_p$ and further is its torsion $\mathbf{Z}_p$-submodule.

The dual of $(\mathbf{Z}_p, +)$ in the category of locally compact $\mathbf{Z}_p$-modules (or $\mathbf{Z}$-modules) is $(\mathbf{T}_p, \times)$. More precisely, if we choose a topological generator a for $(\mathbf{Z}_p, +)$ the mapping $\varphi \longmapsto \varphi(a)$ embeds the 'full (continuous) dual' $\mathrm{Hom}_{\mathbf{Z}}(\mathbf{Z}_p, \mathbf{C}_p^\times)$ in $\mathbf{C}_p^\times$ as $(\mathbf{U}_p, \times)$ and its torsion subgroup (the 'Pontrjagin dual') as $(\mathbf{T}_p, \times)$.

We now choose, once and for all, such a topological generator a for $(\mathbf{Z}_p, +)$ and hence corresponding embeddings of the 'duals' of $(\mathbf{Z}_p, +)$ in $(\mathbf{C}_p^\times, \times)$ (unlike in the 'classical case' over $\mathbf{C}$, the theory which is developed here is partly dependent on the choice of a). For simplicity (and 'without loss of generality') we will suppose that $a = 1$. Thus, explicitly, for each $\beta \in \mathbf{U}_p$, $\varphi_\beta : z \longmapsto \beta^z$ is a continuous character $(\mathbf{Z}_p, +) \longrightarrow (\mathbf{C}_p^\times, \times)$ (and every such character arises in this way).

The purpose of this paper is to survey various attempts to construct a satisfactory analogue of 'Fourier analysis' for spaces of functions on $\mathbf{Z}_p$ and its Pontrjagin dual, the 'p-adic circle' $\mathbf{T}_p$, which take their values in $\mathbf{C}_p$. The main problem here is that, unlike in the 'classical case', $(\mathbf{Z}_p, +)$ does not admit a 'Haar measure' (taking *values* in $\mathbf{C}_p$; indeed the obvious measure to attach to the subgroup $p^n\mathbf{Z}_p$ is $1/p^n$ which tends to infinity p-adically as $n \longrightarrow \infty$ despite the fact that $p^n\mathbf{Z}_p$ itself appears to 'shrink'!). One hopes that this theory might be a special case of a more general 'p-adic harmonic analysis' and, in fact, in his thesis [26] Schikhof has developed such a theory for those zero-dimensional locally compact abelian groups which, together with their duals, *do* admit 'Haar measures' with values in $\mathbf{C}_p$ (see also [6] and [16]).

We now give a brief outline of the contents of the paper (much of the material in Sections 1 to 5 is not new and, in order to keep the length of the paper within bounds, proofs here, for the most part, are omitted).

There are two basic approaches to the problem, one starting with 'summation' on $\mathbf{T}_p$, which is outlined in Section 1, and the other starting with an 'integral' on $\mathbf{Z}_p$, which is outlined in Sections 2, 3, 4. In Section 5 and Section 6 we give some explicit examples and applications of the latter approach (it is difficult to give explicit examples of the former; for an interesting area of application, however, see [15]). These approaches have little overlap and in Section 7 we obtain a natural synthesis of them (essentially by extending the notion of 'tending to zero' on $\mathbf{T}_p$). The main conclusion is that there is a unique (non-invariant) 'integral' on $\mathbf{Z}_p$ which gives rise to a 'satisfactory' Fourier analysis once we have chosen a topological generator for $\mathbf{Z}_p$ as above. Finally, in Section 8, this view is strengthened by considering the action of the Galois group of $\mathbf{C}_p$ over a closed subfield K

of $\mathbb{C}_p$ in the above situation. It turns out that here there is a surprising dichotomy of outcome dependent on whether or not the different of K over $\mathbb{Q}_p$ is zero.

As mentioned above we will have need to consider functions on $\mathbf{T}_p$ which do not tend to zero 'off finite subsets' and so it will be convenient to use the notation $\sum_\omega$ to mean $\lim_{n \longrightarrow \infty} \sum_{\omega \in \mathbf{C}_{p^n}}$ throughout.

The spaces of functions F considered will generally be $\mathbb{C}_p$-Banach spaces or algebras equipped with natural operations and valuations $V: F \longrightarrow \mathbf{R} \cup \{\infty\}$. Note that we choose the *valuation* notation V in preference to the more usual *norm* notation $\| \ \|$ (these being related by $\|f\| = p^{-V(f)}$). We will always use v to denote the 'uniform' valuation on spaces F of functions $f(x)$, that is $v(f) = \inf_x v_p(f(x))$. All mappings considered will be continuous. We denote by F' the dual space of F. For background material on p-adic analysis see for example [2, 8, 18, 19, 23, 25, 27].

Finally, for those readers familiar with the 'hybrid' theory of Fourier analysis for *complex-valued* functions $f: \mathbb{Z}_p \longrightarrow \mathbb{C}$ (as outlined, for example, in [29]; see Chapter 2, Section 6 in the special case $K = \mathbb{Q}_p$), we make a few basic comments by way of comparison with the purely p-adic theory described here. It would be interesting to compare these two theories in more detail.

The two theories have a common algebraic basis; namely in each case the 'locally constant' functions $\mathcal{L}$ on $\mathbb{Z}_p = \varprojlim_n \mathbb{Z}/p^n\mathbb{Z}$ correspond under the Fourier transform (arising from the elementary 'arithmetic mean' integral on $\mathbb{Z}/p^n\mathbb{Z}$) to functions on the discrete Pontrjagin dual of $\mathbb{Z}_p$,

$$C_{p^\infty} = \varinjlim_n C_{p^n},$$

with finite support.

In the hybrid $\mathbb{C}$-valued case this basic algebraic theory for the $\mathbb{Z}/p^n\mathbb{Z}$, C_{p^n} extends 'by continuity' (using the Haar measure, distributions, etc.) to various more substantial spaces of functions on $\mathbb{Z}_p$ in which $\mathcal{L}$ is dense. In particular for L^1-functions we have the Riemann–Lebesgue lemma (the 'Fourier coefficients' tend to zero) and for L^2-functions the Plancherel equality. However, in order to obtain good convergence results for the inversion of Fourier series it is further necessary to choose a 'special ordering' of C_{p^∞} (which thereby gives good properties to the corresponding Dirichlet and Fejer kernels).

On the other hand in the $\mathbb{C}_p$-valued case the analogous theory based on extending the algebraic theory 'by p-adic continuity' still yields only insubstantial spaces of functions on $\mathbb{Z}_p$ (owing to the absence of a p-adic valued Haar measure on $\mathbb{Z}_p$) although the results are analogous. In order to obtain similar results for more substantial spaces of functions the 'arithmetic mean' integral must be more radically extended (essentially uniquely but now non-invariantly) by requiring the persistence of a 'good convolution multiplication' (e.g. associative). However, now the Riemann–Lebesgue lemma fails. It is worth recalling at this point that for a sequence $\{a_n\}$ in $\mathbb{C}_p$, $\sum a_n$ converges precisely when a_n *does* tend to zero and so the archimedean and non-archimedean convergence problems for Fourier inversion here have a rather different flavour. The Riemann–Lebesgue lemma can be recovered in a weak sense (in terms of p-adic analytic continuation) and some convergence results obtained for the inversion of Fourier series (using 'special partial sums' over the C_{p^n} as in the hybrid case) by choosing an embedding of C_{p^∞} as a (discrete) subgroup of $(\mathbb{C}_p^\times, \times)$.

This appears to be the p-adic analogue of the 'special ordering' of C_{p^∞} chosen in the hybrid case; as there it is determined only up to an automorphism of C_{p^∞}. (Note that the corresponding embeddings of C_{p^∞} as a subgroup of $(\mathbb{C}^\times, \times)$ of course do not induce the correct, discrete, topology on C_{p^∞}.) A further new feature in the purely p-adic case arises from the fact that $\mathbb{C}_p$ is not locally compact and the theory becomes substantially simpler when the functions $f\colon \mathbb{Z}_p \longrightarrow \mathbb{C}_p$ considered are restricted so that $f(\mathbb{Z}_p)$ is contained in a fixed finite extension K of $\mathbb{Q}_p$ in $\mathbb{C}_p$ (or more generally if K has 'non-zero different' over $\mathbb{Q}_p$).

1 L^1-Fourier theory

We recall here some basic results from the L^1-Fourier theory due to J. Fresnel and B. de Mathan (see [13, 14], amongst others) which we will need later.

Definition 1.1. We denote by $C(\mathbb{Z}_p)$ the $\mathbb{C}_p$-Banach algebra of all *continuous functions* $f: \mathbb{Z}_p \longrightarrow \mathbb{C}_p$ under the usual pointwise operations and valuation v where $v(f) = \inf_{z \in \mathbb{Z}_p} v_p(f(z))$.

Lemma 1.2. (See [1].) *Let $f, g \in C(\mathbb{Z}_p)$. Then*

1. *There is a unique $f \circledast g \in C(\mathbb{Z}_p)$ with $f \circledast g(n) = \sum_{i=0}^{n} f(i)g(n-i)$ for all $n \in \mathbb{N}$;*
2. *$v(f \circledast g) = v(f) + v(g)$;*
3. *$(C(\mathbb{Z}_p), +, \circledast, v)$ is a commutative $\mathbb{C}_p$-Banach algebra without identity element.*

Theorem 1.3. (See [22] and [1]) *Each $f \in C(\mathbb{Z}_p)$ has a unique uniformly convergent 'Mahler-type' expansion*

$$f = \sum_{n=0}^{\infty} b_n \binom{z+n}{n} = \sum_{n=0}^{\infty} b_n \cdot 1^{\circledast n+1}$$

with $b_n \longrightarrow 0$ and vice versa. *Further, each $b_n = \Delta^n f(-1)$ (where $\Delta g(z) = g(z) - g(z-1)$) and $v(f) = \inf_{n \geqslant 0} v_p(b_n)$.*

Definition 1.4. (See [2].) We denote by $H_0(\mathbf{V}_p)$ the $\mathbb{C}_p$-Banach algebra of all *analytic elements* $h: \mathbf{V}_p \cup \{\infty\} \longrightarrow \mathbb{C}_p$ vanishing at ∞ under the usual pointwise operations and valuation v where $v(h) = \inf_{\alpha \in \mathbf{V}_p} v_p(h(\alpha))$. Recall that if $h: \mathbf{V}_p \cup \{\infty\} \longrightarrow \mathbb{C}_p$ then $h \in H_0(\mathbf{V}_p) \iff h$ is a uniform limit on $\mathbf{V}_p \cup \{\infty\}$ of rational functions vanishing at ∞ and with no pole in $\mathbf{V}_p \cup \{\infty\} \iff h$ has a (unique) uniformly convergent expansion of the form

$$h(\alpha) = \sum_{n=0}^{\infty} b_n (1-\alpha)^{-n-1}$$

with $b_n \longrightarrow 0$ and then $v(h) = \inf_{n \geqslant 0} v_p(b_n)$.

Proposition 1.5. (See [1].)

1. *Let $f \in C(\mathbb{Z}_p)$. Then there is a unique $\theta(f) \in H_0(\mathbf{V}_p)$ with $\sum_{n=0}^{\infty} f(n)\alpha^n$ as its Taylor expansion at 0, viz.*

$$\theta(f)(\alpha) = \sum_{n=0}^{\infty} b_n (1-\alpha)^{-n-1}$$

(with b_n as in Theorem 1.3).

2. $\theta: (C(\mathbb{Z}_p), \circledast, v) \longrightarrow (H_0(\mathbf{V}_p), \times, v)$ *is an isometric isomorphism of* $\mathbb{C}_p$*-Banach algebras.*

Definition 1.6. (See [14].) We denote by $L^1(\mathbf{T}_p)$ the $\mathbb{C}_p$-Banach algebra of all functions $h: \mathbf{T}_p \longrightarrow \mathbb{C}_p$ with $\lim_\omega h(\omega) = 0$ (limit 'off finite subsets of $\mathbf{T}_p$') under usual pointwise addition, convolution multiplication $*$ where

$$g * h(\omega) = \sum_\sigma g(\omega\sigma^{-1})h(\sigma),$$

and valuation v where

$$v(h) = \inf_\omega v_p(h(\omega))$$

(taken over $\omega \in \mathbf{T}_p$).

Definition 1.7. (See [14].) We define the *Fourier transform*

$$\mathcal{F}: (L^1(\mathbf{T}_p), *, v) \longrightarrow (C(\mathbb{Z}_p), \times, v)$$

by putting, for all $g \in L^1(\mathbf{T}_p)$,

$$\mathcal{F}(g) = \sum_\omega g(\omega)\varphi_{\omega^{-1}} \in C(\mathbb{Z}_p)$$

(a uniformly convergent 'series'; recall that $\varphi_\beta(z) = \beta^z$). Clearly $v(\mathcal{F}(g)) \geqslant v(g)$ and $\mathcal{F}$ is a continuous homomorphism of $\mathbb{C}_p$-Banach algebras.

Theorem 1.8. (See [13] and [14].)

1. $\mathcal{F}$ *is surjective but not injective.*

2. $\ker \mathcal{F}$ *consists of all* $g \in L^1(\mathbf{T}_p)$ *such that there exists* $\{b_n\}$ *with* $v_p(b_n) + \log_p(n) \longrightarrow \infty$ *and* $g(\omega) = \sum_{n=0}^{\infty} b_n(\omega - 1)^n$ *for all* $\omega \in \mathbf{T}_p$.

3. $\mathcal{F}$ *induces an isometric isomorphism of* $\mathbb{C}_p$*-Banach algebras:*

$$\overline{\mathcal{F}}: (L^1(\mathbf{T}_p)/\ker(\mathcal{F}), *, \bar{v}) \longrightarrow (C(\mathbb{Z}_p), \times, v).$$

Remark 1.9. The non-injectivity of $\mathcal{F}$ has the following interpretation in terms of analytic continuation on unpleasant sets (see [4]): If $g \in L^1(\mathbf{T}_p)$ then $\alpha \longmapsto \sum_{n=0}^{\infty} \mathcal{F}(g)(n)\alpha^n$ for $(v_p(\alpha) > 0)$ extends uniquely to an analytic element in $H_0(\mathbf{V}_p)$ (see 1.5) $\alpha \longmapsto \theta(\mathcal{F}(g))(\alpha)$ which in turn extends *non-uniquely* in general to an analytic element $\alpha \longmapsto \sum_\omega g(\omega)(1 - \omega^{-1}\alpha)^{-1}$ on any subset of $\mathbb{C}_p \cup \{\infty\}$ bounded away from $\mathbf{T}_p$.

Remark 1.10. Unfortunately $\mathcal{F}$ has no 'section' $\Lambda: C(\mathbb{Z}_p) \longrightarrow L^1(\mathbf{T}_p)$ which is a continuous homomorphism of $\mathbb{C}_p$-Banach algebras. Further, there does not appear to be a section of $\mathcal{F}$ on a 'big' subspace of $C(\mathbb{Z}_p)$ (for example, containing the polynomial functions) given by a 'natural' formula (see also Section 4; of course if $f \in C(\mathbb{Z}_p)$ is constant on cosets of $p^N\mathbb{Z}_p$ in $\mathbb{Z}_p$ then $f = \sum_{\omega \in C_{p^N}} a_\omega \cdot \varphi_{\omega^{-1}}$ where $a_\omega = \sum_{i=0}^{p^N-1} f(i)\omega^i/p^N$.)

We will also need the following result which arises in the course of the proof of Theorem 1.8:

Lemma 1.11. (See [14].) *Let* $M, N > 0$. *Then there exists* $g \in L^1(\mathbf{T}_p)$ *with* $g(\omega) = 0$ *for all* $\omega \in C_{p^M}$, $v(g) \geqslant -1/N$ *and* $\mathcal{F}(g) = 1$.

Definition 1.12. (See [1, 3].) We denote by $M(\mathbf{Z}_p) = C(\mathbf{Z}_p)'$ the $\mathbf{C}_p$-Banach space of all *measures* $\mu: C(\mathbf{Z}_p) \longrightarrow \mathbf{C}_p$ under the valuation v where $v(\mu) = \inf v_p(\mu(f))$ (the infimum being taken over all $f \in C(\mathbf{Z}_p)$ with $v(f) = 0$).

Lemma 1.13. (See [1, 3].) *Each measure* $\mu \in M(\mathbf{Z}_p)$ *corresponds to a unique bounded sequence* $\{b_n\}$ $(n \geqslant 0)$ *in* $\mathbf{C}_p$ *such that for all* $\beta \in \mathbf{U}_p$,

$$\mu(\varphi_\beta) = \sum_{n=0}^{\infty} b_n(\beta - 1)^n$$

(and 'vice versa'). Furthermore we have,

$$b_n = \mu(\binom{z}{n})$$

and

$$v(\mu) = \inf_{n \geqslant 0} v_p(b_n).$$

Lemma 1.14. (With the notation of 1.13.) *For all* $M \geqslant 0$,

$$v(\mu) = \inf_{\omega \in \mathbf{T}_p} v_p(\mu(\varphi_\omega)) = \inf v_p(\mu(\varphi_\omega))$$

taken over $\omega \in \mathbf{T}_p \backslash C_{p^M}$.

Proof This follows easily from Theorem 1.8(3) and Lemma 1.11. ◊ ◊

Remark 1.15. Thus $M(\mathbf{Z}_p)$ with the appropriate definition of convolution multiplication (see [1, 3]) can be naturally identified with the algebra of bounded analytic functions on $\mathbf{U}_p$ and each such function is determined by its asymptotic behaviour on $\mathbf{T}_p$.

2 Integrals

We now turn to consider the other basic approach to Fourier series via 'integrals' on $\mathbb{Z}_p$:

Definition 2.1. We denote by $\mathrm{UD}(\mathbb{Z}_p)$ the $\mathbb{C}_p$-Banach algebra of all *uniformly differentiable* functions $f: \mathbb{Z}_p \longrightarrow \mathbb{C}_p$ under the usual pointwise operations and valuation V where $V(f) = \min\{v(f), R(f)\}$ with

$$R(f) = \inf\left\{v_p\left(\frac{f(x)-f(y)}{x-y}\right) : x, y \in \mathbb{Z}_p,\, x \neq y \in \mathbb{Z}_p\right\}.$$

Remark 2.2. Recall that $f: \mathbb{Z}_p \longrightarrow \mathbb{C}_p$ is said to be *uniformly* (or *strictly*) *differentiable* if the difference quotient

$$(x, y) \longmapsto \frac{f(x)-f(y)}{x-y} \quad \text{for } x \neq y$$

extends to a (necessarily unique) continuous function on $\mathbb{Z}_p \times \mathbb{Z}_p$ to $\mathbb{C}_p$. This is a much stronger and more amenable condition than '*continuously differentiable*' in the p-adic case. For example, if $f \in C(\mathbb{Z}_p)$ then, with the notation of Theorem 1.3, $f \in \mathrm{UD}(\mathbb{Z}_p) \iff v_p(b_n) - \log_p(n) \longrightarrow \infty$ as $n \longrightarrow \infty$. (There seems to be no such simple criterion in the case of continuously differentiable functions; see [5, 17, 23] or [32].)

Definition 2.3. (See [7], [30] and [35].) We denote by $\mathrm{Int}\,(\mathbb{Z}_\mathrm{p})$ the closed $\mathbb{C}_p$-subalgebra of $\mathrm{UD}(\mathbb{Z}_p)$ consisting of all *Riemann integrable* functions $f: \mathbb{Z}_p \longrightarrow \mathbb{C}_p$ so that

$$\mathrm{Int}\,(\mathbb{Z}_\mathrm{p}) = \{\mathrm{f} \in \mathrm{UD}(\mathbb{Z}_\mathrm{p}) | \mathrm{f}' = 0\}.$$

We denote by $I \in \mathrm{Int}\,(\mathbb{Z}_\mathrm{p})'$ the *Riemann integral* on $\mathrm{Int}\,(\mathbb{Z}_\mathrm{p})$ (in the obvious sense).

Remark 2.4. Let $\chi_{x,k}$ denote the characteristic function of $x + p^k\mathbb{Z}_p$ on $\mathbb{Z}_p$ ($x \in \mathbb{Z}_p$, $k \geqslant 0$). Then $I(\chi_{x,k}) = p^{-k}$ and the $\chi_{x,k}$ have dense linear span in $\mathrm{Int}\,(\mathbb{Z}_\mathrm{p})$ (and also in $C(\mathbb{Z}_p)$ but not in $\mathrm{UD}(\mathbb{Z}_p)$!).

Remark 2.5. $\mathrm{Int}\,(\mathbb{Z}_\mathrm{p})$ is the natural home for 'Haar measure' on $\mathbb{Z}_p$ (I is an 'essentially unique' translation invariant integral on $\mathrm{Int}\,(\mathbb{Z}_\mathrm{p})$) and there is a satisfactory analogue of classical Fourier theory for $\mathrm{Int}\,(\mathbb{Z}_\mathrm{p})$ based on I (see [34] and [35]). However, unfortunately $\mathrm{Int}\,(\mathbb{Z}_\mathrm{p})$ is lacking in 'interesting' functions and no 'reasonable' (for example, containing the polynomials as a dense subspace) $\mathbb{C}_p$-Banach space of functions on $\mathbb{Z}_p$ admits a non-zero translation-invariant integral.

Definition 2.6. $J \in \mathrm{UD}(\mathbb{Z}_p)'$ is said to be an *integral* if J extends the Riemann integral $I \in \mathrm{Int}\,(\mathbb{Z}_\mathrm{p})'$ (so J is an ersatz 'Haar integral' on $\mathbb{Z}_p$).

Remark 2.7. Despite the absence of a Hahn-Banach theorem in this context there are many (non-invariant) integrals $J \in \mathrm{UD}(\mathbb{Z}_p)'$. In fact differentiation D induces a $\mathbb{C}_p$-linear isometry of $\mathrm{UD}(\mathbb{Z}_p)/\mathrm{Int}\,(\mathbb{Z}_p)$ onto $C(\mathbb{Z}_p)$ and so clearly the set of all such integrals J is a coset of $M(\mathbb{Z}_p) \circ D$ in $\mathrm{UD}(\mathbb{Z}_p)'$, '*the Haar coset*' (see [11], [35] and [36]).

Examples 2.8. We now give two natural methods for constructing integrals $J \in \mathrm{UD}(\mathbb{Z}_p)'$. (It would be interesting to find others.)

(i) $I_{a,b}$ *arising from special Riemann sums* (see [9, 20, 31, 33] and [8], Chapter 1, Section 3, for a historical comment): Let $a \in \mathbb{Z}_p \backslash p\mathbb{Z}_p$ and $b \in \mathbb{Z}_p$. For each $f \in \mathrm{UD}(\mathbb{Z}_p)$,

$$I_{a,b}(f) = \lim_{n \longrightarrow \infty} \sum_{i=0}^{p^n-1} f(ai+b)/p^n$$

exists and $I_{a,b} \in \mathrm{UD}(\mathbb{Z}_p)'$ is an integral. Clearly the $I_{a,b}$ are all related by 'affine changes of variable on $\mathbb{Z}_p$'. $I_{1,0}$ will henceforth be denoted by I (this will turn out to be our preferred integral!).

(ii) *Relations between the* $I_{a,b}$ (see the remark following 2.6):
(a) $I_{1,0}(f) - I_{a,0}(f) = \mu_a(f')$ where $\mu_a \in M(\mathbb{Z}_p)$ is a *Mazur measure* (which, together with $I_{1,0}$, arises in the theory of p-adic L-functions; see [18, 19] and [20]).
(b) $I_{1,b}(f) - I_{1,0}(f) = f' \circledast 1(b-1)$. This can be used to evaluate certain integrals with ease. For example, putting $b = 1$ and $f(x) = \exp(\alpha x)$ where $v_p(\alpha) > \frac{1}{p-1}$ we immediately obtain $I(\exp(\alpha x)) = \alpha/(\exp(\alpha) - 1)$ and so (on equating Taylor series) $I(x^n) = B_n$ for all $n \geqslant 0$ where B_n denotes the nth Bernoulli number (see [20, 33]).

(iii) I_P *arising from a 'primitive'* P: There exist (many) $\mathbb{C}_p$-linear isometries $P: C(\mathbb{Z}_p) \longrightarrow \mathrm{UD}(\mathbb{Z}_p)$ such that $P(f)' = f$ for all $f \in C(\mathbb{Z}_p)$. (For example, $P(f)(x) = \sum_{n=0}^{\infty}(x_{n+1} - x_n)f(x_n)$ where, for each $x \in \mathbb{Z}_p$ and $n \geqslant 0$, $x_n \in \mathbb{N}$ with $0 \leqslant x_n < p^n$ and $v_p(x - x_n) \geqslant n$.) If P is such a 'primitive' then we put $I_P(f) = I(f - P(f'))$ for all $f \in \mathrm{UD}(\mathbb{Z}_p)$ (note that $f - P(f') \in \mathrm{Int}\,(\mathbb{Z}_p)$). Then $I_P \in \mathrm{UD}(\mathbb{Z}_p)'$ is an integral (see [11, 27]).

3 Convolutions

Definition 3.1. (See [35] and [36].) Let $J \in \mathrm{UD}(\mathbb{Z}_p)'$ be an integral and let $f, g \in \mathrm{UD}(\mathbb{Z}_p)$. Then we define the J-convolution $f *_J g \in \mathrm{UD}(\mathbb{Z}_p)$ by putting, for all $z \in \mathbb{Z}_p$,

$$f *_J g(z) = \lim_{n \longrightarrow \infty} p^n J^{(x)} J^{(y)}(f(x)g(y)\chi_{z,n}(x+y))$$

where $J^{(t)}$ denotes J-integration with respect to the t-variable and $\chi_{z,n}$ denotes the characteristic function of $z + p^n \mathbb{Z}_p$ on $\mathbb{Z}_p$.

Remark 3.2. If we were to define $f *_J g$ in the obvious way by putting

$$f *_J g(z) = J^{(x)}(f(x)g(z-x)),$$

then unfortunately $f *_J g \notin \mathrm{UD}(\mathbb{Z}_p)$ and $f *_J g \neq g *_J f$ in general. In fact since J is not translation invariant there is no natural way to apply it to functions on $Y = \{(x,y) \in \mathbb{Z}_p \times \mathbb{Z}_p | x + y = z\}$. Instead we consider a limit of 'normalized' repeated J-integrals on $\mathbb{Z}_p \times \mathbb{Z}_p$ restricted to gradually shrinking neighbourhoods of Y in $\mathbb{Z}_p \times \mathbb{Z}_p$ (which of course reduces to the usual result in the 'classical' situation).

Remark 3.3. In fact $f *_J g = g *_J f$ but in general $*_J$ is not associative.

Theorem 3.4. (See [36].) *The following conditions are equivalent:*

1. $*_J$ *is associative,*
2. $J(f *_J g) = J(f)J(g)$ *for all* $f, g \in \mathrm{UD}(\mathbb{Z}_p)$,
3. $J = I_{a,0}$ *for some* $a \in \mathbb{Z}_p \backslash p\mathbb{Z}_p$, *that is* J *is a dilation of* $I = I_{1,0}$.

Remark 3.5. Thus, surprisingly, we have a purely analytic reason for singling out these integrals which arise in the arithmetic theory of p-adic L-functions (see [18, 19] and [20]).

Remark 3.6. Note that $I_{1,0} = I$ and $I_{1,1} = I_{-1,0}$ but no other *translation* of I satisfies the equivalent conditions of Theorem 3.4.

For the rest of this section we will consider the case $J = I$ in more detail and put $* = *_I$:

Theorem 3.7. (See [35].)

1. $(\mathrm{UD}(\mathbb{Z}_p), +, *, V)$ *is a commutative* $\mathbb{C}_p$*-Banach algebra without identity element (in the slightly weakened sense that* $V(f * g) \geqslant V(f) + V(g) - 1$ *for all* $f, g \in \mathrm{UD}(\mathbb{Z}_p)$*).*

2. *Differentiation* $D: (\mathrm{UD}(\mathbb{Z}_p), *, V) \longrightarrow (C(\mathbb{Z}_p), -\circledast, v)$ *is a surjective continuous homomorphism of* $\mathbb{C}_p$*-Banach algebras with kernel* $\mathrm{Int}(\mathbb{Z}_p)$. *Further* D *induces an isometric isomorphism of* $\mathbb{C}_p$*-Banach algebras*

$$\overline{D}: (\mathrm{UD}(\mathbb{Z}_p)/\mathrm{Int}(\mathbb{Z}_p), *, \overline{\mathrm{V}}) \longrightarrow (\mathrm{C}(\mathbb{Z}_p), -\circledast, \mathrm{v}).$$

Remark 3.8. The convolution multiplication $*$ on $\mathrm{UD}(\mathbb{Z}_p)$ does not of course behave quite in the classical way. In fact, for all $z \in \mathbb{Z}_p$,

$$f * g(z) = I^{(x)}(f(x)g(z-x)) - f \circledast g'(z)$$

(note that in general neither term on the right-hand side is in $\mathrm{UD}(\mathbb{Z}_p)$). The multiplications $\times$, $*$, $\circledast$ and the integral I on $\mathrm{UD}(\mathbb{Z}_p)$ have many other interrelations of a 'cohomological character'. For example, for all $f, g \in \mathrm{UD}(\mathbb{Z}_p)$, we have (see [35]):

1. $(f * g)' = -f' \circledast g'$;
2. $f * g = (f \circledast g)' - f' \circledast g - f \circledast g'$;
3. $z(f \circledast g) = zf \circledast g + f \circledast zg$;
4. $f \circledast g = z(f * g) - zf * g - f * zg$.

We now briefly recall some results from the 'Gelfand Theory' of the algebra $(\mathrm{UD}(\mathbb{Z}_p), *, V)$ (see [21] for the 'classical' theory and [12] for some special problems associated with the p-adic case):

Definition 3.9. (See [36].) Let $X_p = \mathbf{V}_p \cup \{\infty\} \cup \mathbf{T}_p$ and let $C_0(X_p)$ denote the $\mathbb{C}_p$-Banach algebra of all continuous functions $h: X_p \longrightarrow \mathbb{C}_p$ vanishing at ∞ under the usual pointwise operations and valuation v where $v(h) = \inf v_p(h(x))$ taken over $x \in X_p$.

Then the *Gelfand transform* $\mathcal{G}: \mathrm{UD}(\mathbb{Z}_p) \longrightarrow C_0(X_p)$ is defined by putting, for all $f \in \mathrm{UD}(\mathbb{Z}_p)$,

$$\left.\begin{array}{rcl} \mathcal{G}(f)(\omega) & = & I(f\varphi_\omega) \text{ for } \omega \in \mathbf{T}_p \\ \mathcal{G}(f)(\alpha) & = & -\theta(f')(\alpha) \text{ for } \alpha \in \mathbf{V}_p \cup \{\infty\} \end{array}\right\}$$

so that $\mathcal{G}(f) \in C_0(X_p)$.

Remark 3.10. The restriction of $\mathcal{G}(f)$ to $\mathbf{V}_p \cup \{\infty\}$ of course belongs to $H_0(\mathbf{V}_p)$. We will have more to say about $\mathrm{Im}(\mathcal{G})$ in Section 7.

Theorem 3.11. (See [36].)

1. *Let* W *be the spectral valuation on* $(\mathrm{UD}(\mathbb{Z}_p), *, V)$ *defined so that, for all* $f \in \mathrm{UD}(\mathbb{Z}_p)$, $W(f) = \lim_{n \longrightarrow \infty} V(f^{*n})/n$ *and* $W(f) \geqslant V(f) - 1$. *Then* $(\mathrm{UD}(\mathbb{Z}_p), *, W)$ *is a non-complete normed* $\mathbb{C}_p$*-algebra and the Gelfand transform*

$$\mathcal{G}: (\mathrm{UD}(\mathbb{Z}_p), *, W) \longrightarrow (C_0(X_p), \times, v)$$

is a (non-surjective) isometric homomorphism of normed $\mathbb{C}_p$*-algebras.*

2. Thus for all $f \in \mathrm{UD}(\mathbb{Z}_p)$, $W(f) = v(\mathcal{G}(f))$.

More precisely:

(a) $W(f) = \inf_\omega v_p(\mathcal{G}(f)(\omega))$ *taken over* $\omega \in \mathbf{T}_p$,

(b) and

$$\begin{aligned} v(f') &= \lim_{\omega} v_p(\mathcal{G}(f)(\omega)) \text{ off finite subsets of } \mathbf{T}_p \\ &= \inf_{\alpha} v_p(\mathcal{G}(f)(\alpha)) \text{ taken over } \alpha \in \mathbf{V}_p \cup \{\infty\}. \end{aligned}$$

(For a more precise result see [36], Theorem 4.8.)

Example 3.12. For $n > 0$, $V(z^n) = 0$ but $W(z^n) = -\overline{n}/(p-1)$ where $1 \leqslant \overline{n} \leqslant p-1$ and $n \equiv \overline{n} \bmod p-1$. (See [37].)

4 J-Fourier theory

Let $J \in \mathrm{UD}(\mathbb{Z}_p)'$ be a fixed integral. We now briefly describe the associated 'J-Fourier theory' which, unfortunately, has little overlap with the L^1-Fourier theory described in Section 1 (see Theorem 1 below):

Theorem 4.1. (See [36].) *For all $f \in \mathrm{UD}(\mathbb{Z}_p)$ and $\omega \in \mathbf{T}_p$ put $\hat{f}(\omega) = J(f\varphi_\omega)$. Then:*

- Riemann–Lebesgue lemma: $\lim_\omega \hat{f}(\omega) = 0 \iff f' = 0$ *(that is $f \in$* Int $(\mathbb{Z}_p)$ *is* Riemann integrable*; this follows by an argument similar to that employed in Remark* 7.6*).*
- Fourier inversion formula*: $f = \sum_\omega \hat{f}(\omega)\varphi_{\omega^{-1}}$ (where convergence is in the uniform v-valuation but not in the V-valuation unless $f' = 0$).*
- Parseval's formula*: $J(fg) = \sum_\omega \hat{f}(\omega)\hat{g}(\omega^{-1})$ for all $f, g \in \mathrm{UD}(\mathbb{Z}_p)$.*
- Convolution formula*: $f *_J g = \sum_\omega \hat{f}(\omega)\hat{g}(\omega)\varphi_{\omega^{-1}}$ for all $f, g \in \mathrm{UD}(\mathbb{Z}_p)$ (with uniform convergence).*

Remark 4.2. Note that in Theorem 2, $\sum_\omega$ denotes

$$\lim_{n \longrightarrow \infty} \sum_{\omega \in C_{p^n}}$$

and

$$\sum_\omega \varphi_\omega = 0,$$

a lack of uniqueness for 'Fourier-type' series which can cause problems!

Remark 4.3. In Theorem 4, the right-hand side of the convolution formula is *not* in general the J-Fourier series of the left-hand side of the equation. In fact this is always the case precisely when J satisfies the equivalent conditions of Theorem 3.4, that is J is a dilation of I. We will see later (in Section 7) that only the I-Fourier coefficients $\hat{f}(\omega)$ have 'good' asymptotic behaviour.

For the rest of the article, unless stated otherwise, we will suppose that $J = I$ and so clearly $\hat{f}(\omega) = \mathcal{G}(f)(\omega)$ where $\mathcal{G}(f)$ denotes the Gelfand transform of f (see Section 3).

5 Examples

5.1 I-Fourier series expansions

For further details on this material, see [33, 35].

(a) For $z \in \mathbb{Z}_p$, we have the uniformly convergent series

$$z = -\frac{1}{2} + \sum_{\omega \neq 1} (\omega - 1)^{-1} \omega^{-z}.$$

As we have seen in Section 2, $I(z^n) = B_n$ for all $n \geqslant 0$; see [33] for the complete I-Fourier series expansion of z^n.

(b) For $\beta \in \mathsf{U}_p \backslash \mathsf{T}_p$,

$$\beta^z = \sum_{\omega} (\beta\omega - 1)^{-1} \log \beta \cdot \omega^{-z}$$

by a similar calculation to that in 2.8(ii)(b).

5.2 I-Convolutions

For further details on this material, see [35].

(a) We have

$$z * z = -\frac{1}{6} - z - \frac{z^2}{2} = \frac{1}{4} + \sum_{\omega \neq 1} (\omega - 1)^{-2} \omega^{-z}.$$

The various interrelations between $\times$, $*$, $\circledast$, I mentioned in Section 3 can be used to evaluate $z^m * z^n$ etc. by recursion.

(b) We have

$$\begin{aligned} \gamma^z * \beta^z &= (\beta - \gamma)^{-1} \{\log \beta \cdot \gamma^{z+1} - \log \gamma \cdot \beta^{z+1}\} \\ &= \sum_{\omega} (\gamma\omega - 1)^{-1} \cdot (\beta\omega - 1)^{-1} \cdot \log \gamma \cdot \log \beta \cdot \omega^{-z} \end{aligned}$$

where $\gamma, \beta \in \mathsf{U}_p \backslash \mathsf{T}_p, \gamma \neq \beta$.

5.3 Parseval's formula

See [34, 35] for further details.

Let $f \in \mathrm{UD}(\mathbb{Z}_p)$ and $\beta \in \mathbb{U}_p \backslash \mathbb{T}_p$. Then by 2 and 3 we have

$$I(f\varphi_\beta) = \sum_\omega \hat{f}(\omega^{-1}) \cdot (\beta\omega - 1)^{-1} \cdot \log \beta.$$

On the other hand we clearly have that, for all $\beta \in \mathbb{U}_p$,

$$I(f\varphi_\beta) = \sum_{n=0}^{\infty} I\left(f \cdot \binom{z}{n}\right)(\beta - 1)^n$$

which represents an analytic function of $\beta \in \mathbb{U}_p$ (in fact

$$v_p\left(I\left(f \cdot \binom{z}{n}\right)\right) + \log_p n$$

is bounded below for $n \geqslant 0$).

Thus the *restriction* of the Gelfand transform $\mathcal{G}(f)$ of f from X_p to $\mathbb{T}_p$ now *extends* again by *interpolation* to an analytic function $\hat{f}$ on $\mathbb{U}_p$. (Recall that $\mathbb{U}_p$ is the full p-adic dual of $\mathbb{Z}_p$ while $\mathbb{T}_p$ is its 'Pontrjagin' dual. Note also that the convolutions $*$ and $\circledast$ arise naturally through the 'division algorithm' $\hat{f}\hat{g} = (f \hat{*} g) - \log\alpha \cdot (f \hat{\circledast} g)$ on $\mathbb{U}_p$; see [34] for a further development of this approach to Fourier analysis and interpolation.)

In fact P. Cassou-Noguès has shown that these functions are related in that

$$I(f\varphi_\beta) = -\log\beta \cdot \theta(f)(\beta) - \theta(f')(\beta)$$

holds as a *formal* identity (in an appropriate sense) and if f is actually locally analytic then, by analytic continuation, as an *analytic function* identity in a suitable annulus (see [9, 10]).

5.4 Valuation estimates

For reasons of space we have omitted much mention of valuation estimates. For example, for all $f \in \mathrm{UD}(\mathbb{Z}_p)$, we have

$$v_p(I(f)) \geqslant W(f) \geqslant V(f) - 1,$$

and there are numerous other identities and estimates involving v_p, v, V, W, I, $*$, $\circledast$, $\times$, etc. (see [33, 34, 35, 36, 37, 38]). In particular since $I(z^n) = B_n$ we can easily recover the 'Iwasawa algebra' congruences for the modified B_n (which, of course, can be equivalently obtained by using Mazur measures etc.) but also apparently new congruences for the $A^*_{m,n} = I(z^m \circledast z^n)$ ('two variable Bernoulli numbers') which are related to a 'two-variable Riemann zeta function' $Y(w, z)$ discussed in [38] (see also [28]).

6 Difference equations

We first summarize some basic facts concerning difference equations on $\mathbb{Z}_p$ which we will need (see [24] for a general treatment):

Recap 6.1. Define the translation operator $T\colon \mathrm{UD}(\mathbb{Z}_p) \longrightarrow \mathrm{UD}(\mathbb{Z}_p)$ by putting $Tf(z) = f(z-1)$ for all $f \in \mathrm{UD}(\mathbb{Z}_p)$ and $z \in \mathbb{Z}_p$. Let $P(t) = \sum_{r=0}^{k} p_r \cdot t^r \in \mathbb{C}_p[t]$ with $\deg(P(t)) = k \geqslant 0$. Suppose that

$$P(t) = \prod_{i=1}^{r}(t-\beta_i)^{m_i} \cdot R(t)$$

where $\beta_1, \beta_2, \ldots, \beta_r$ are the distinct roots of $P(t)$ in $\mathbb{U}_p$, each $m_i \geqslant 1$ and $R(t) \in \mathbb{C}_p[t]$ has no root in $\mathbb{U}_p$. Put $d = \sum_{i=1}^{r} m_i$. Then:

1. $P(T)\colon \mathrm{UD}(\mathbb{Z}_p) \longrightarrow \mathrm{UD}(\mathbb{Z}_p)$ is surjective.
2. $\ker(P(T))$ has dimension d as a vector space over $\mathbb{C}_p$ and has as a basis the exponential polynomials $z^m\beta_i^{-z}$ with $1 \leqslant i \leqslant r$ and $0 \leqslant m < m_i$.
3. Let $D\colon \ker(P(T)) \longrightarrow \ker(P(T))$ denote differentiation. Then $\ker(D)$ has as a basis the exponential functions β_i^{-z} with $1 \leqslant i \leqslant r$ and $\beta_i \in \mathbb{T}_p$.

Hence, D is an isomorphism $\Longleftrightarrow$ all the $\beta_i \in \mathbb{U}_p \backslash \mathbb{T}_p$ for $1 \leqslant i \leqslant r$.

Remark 6.2. Part 1. above follows easily from the corresponding result for $C(\mathbb{Z}_p)$ obtained in [24] on using the 'Mahler coefficient' characterization of $\mathrm{UD}(\mathbb{Z}_p)$ given in the remark following Definition 2.1.

Lemma 6.3. (Retaining the notation of 6.1.) *Let $f, g \in \mathrm{UD}(\mathbb{Z}_p)$ with $P(T)(g) = f$. Then, for all $\omega \in \mathbb{T}_p$,*

$$P(\omega)\hat{g}(\omega) = \hat{f}(\omega) + \sum_{j=1}^{k} g'(-j) \sum_{r=j}^{k} p_r \cdot \omega^{r-j}.$$

Proof By 2.8(ii)(b) with $b = 1$ we obtain $I(h) - I(T(h)) = h'(-1)$ for all $h \in \mathrm{UD}(\mathbb{Z}_p)$. The result now follows easily by repeated application. ◊ ◊

Remark 6.4. For simplicity we now restrict attention to the difference equation $P(T)(g) = f$ in the case when the $g'(-j)$ $(1 \leqslant j \leqslant k)$ are freely at our disposal. (This is the only case that we will need; the general case needs more careful consideration. Note that $\sum_\omega \varphi_\omega = 0$ so that apparently different solutions given by 'Fourier-type' series may in fact be the same.) From 6.1 this means that $d = k$ and no $\beta_i \in \mathbb{T}_p$, that is

© all the roots of $P(t)$ are in $\mathsf{U}_p \backslash \mathsf{T}_p$.

In this case, by 6.3 (since no $P(\omega) = 0$) we can uniquely solve the difference equation $P(T)(g) = f$ 'explicitly' (as an I-Fourier series expansion) for g in terms of f subject to the 'differential constraints' $g'(-j) = a_j$ $(1 \leqslant j \leqslant k)$ for any choice of the a_j. (Compare this with the usual problem with 'initial conditions' $g(-j) = a_j$.)

We will need the following result in Section 7:

Theorem 6.5. (Retaining the notation of 6.1 and 6.3.) *The rational functions with no pole on X_p and vanishing at ∞ are precisely the Gelfand transforms of the exponential polynomials on $\mathbb{Z}_p$ of the form $\sum_{i=1}^{r} q_i(z)\beta_i^{-z}$ where the $q_i(z) \in \mathbb{C}_p[z]$ and the $\beta_i \in \mathsf{U}_p \backslash \mathsf{T}_p$.*

Proof Suppose that $P(t) \in \mathbb{C}_p[t]$ is as in 6.1 and 6.3 and satisfies condition © above. Then taking $f = 0$ we see that the homogeneous difference equation $P(T)(g) = 0$ has a k-dimensional $\mathbb{C}_p$-vector space of exponential polynomial solutions of the form

$$g(z) = \sum_{i=1}^{r} q_i(z)\beta_i^{-z}$$

where $\deg(q_i(z)) < m_i$. On the other hand these solutions have I-Fourier coefficients of the form

$$\hat{g}(\omega) = \sum_{j=1}^{k} a_j \sum_{r=j}^{k} p_r \cdot \omega^{r-j}/P(\omega), \qquad (\omega \in \mathsf{T}_p)$$

where the a_j $(1 \leqslant j \leqslant k)$ are arbitrary (and $g'(-j) = a_j$). Thus

$$\hat{g}(\omega) = \frac{Q(\omega)}{P(\omega)}$$

where $Q(t) \in \mathbb{C}_p[t]$ with $\deg(Q(t)) < \deg(P(t)) = k$. Clearly by suitable choice of the a_j each such $Q(t)$ arises (exactly once).

Now for $\alpha \in \mathsf{V}_p \cup \{\infty\}$ with $v_p(\alpha) > 0$ we have

$$\mathcal{G}(g)(\alpha) = -\theta(g')(\alpha) = -\sum_{n=0}^{\infty} g'(n)\alpha^n$$

and so since $P(T)(g') = 0$ also a direct calculation shows that

$$P(\alpha)\mathcal{G}(g)(\alpha) = Q(\alpha).$$

Hence by analytic continuation, we have

$$\mathcal{G}(g)(\alpha) = \frac{Q(\alpha)}{P(\alpha)}$$

for all $\alpha \in \mathbf{V}_p \cup \{\infty\}$ (note that both sides are in $H_0(\mathbf{V}_p)$). Thus the Gelfand transform

$$\mathcal{G}(g)(\alpha) = \frac{Q(\alpha)}{P(\alpha)}$$

for all $\alpha \in X_p$. The result now follows easily. ◊ ◊

7 H_0-Fourier theory

We now describe a natural approach to Fourier theory which turns out in some sense to be the amalgam of the L^1 and I theories. This has the advantage that the Fourier transform is defined by 'summation' on $\mathbf{T}_p$ and 'integration' on $\mathbb{Z}_p$ on substantial parts of the appropriate function spaces (see also Section 8).

Definition 7.1. We will denote by $H_0(X_p)$ the $\mathbf{C}_p$-Banach algebra of all *analytic elements* $h: X_p \longrightarrow \mathbf{C}_p$ vanishing at ∞ under the usual pointwise operations and valuation v where $v(h) = \inf v_p(h(\alpha))$ taken over $\alpha \in X_p$.

Recall that if $h: X_p \longrightarrow \mathbf{C}_p$ then $h \in H_0(X_p) \Longleftrightarrow h$ is a uniform limit on X_p of rational functions vanishing at ∞ and with no pole in X_p.

Remark 7.2. Clearly $H_0(X_p)$ is a closed subalgebra of $C_0(X_p)$ (see 3.9).

Lemma 7.3. *For all* $h \in H_0(X_p)$,

1. $v(h) = \inf v_p(h(\omega))$ *taken over* $\omega \in \mathbf{T}_p$.
2. $\lim_\omega v_p(h(\omega)) = \inf v_p(h(\alpha))$ *taken over* $\alpha \in \mathbf{V}_p \cup \{\infty\}$.

Proof It is clearly enough to prove the result if h is a rational function vanishing at ∞ and with no pole in X_p. However, by 6.5, h is the Gelfand transform $\mathcal{G}(f)$ of some exponential polynomial $f \in \mathrm{UD}(\mathbb{Z}_p)$. The result now follows using Theorem 3.11(2). ◊ ◊

Remark 7.4. By Lemma 7.3 it is clear that $H_0(X_p)$ may be naturally identified with its restriction to $\mathbf{T}_p$ and so viewed as a $\mathbf{C}_p$-Banach algebra of functions on $\mathbf{T}_p$. Indeed,

$$H_0(X_p) \cong \left\{ h: \mathbf{T}_p \longrightarrow \mathbf{C}_p \;\middle|\; \begin{array}{l} h \text{ is a uniform limit on } \mathbf{T}_p \text{ of} \\ \text{rational functions with poles in} \\ \mathbf{U}_p \backslash \mathbf{T}_p \text{ and vanishing at } \infty. \end{array} \right\}$$

From this point of view we can think of the condition on h as being of the form '$\lim_\omega h(\omega) = 0$ weakly' which, as we will see in Theorem 7.5, includes the case of $\lim_\omega h(\omega) = 0$ in the usual sense. However, unlike $L^1(\mathbf{T}_p)$ this space is *not* invariant under automorphisms of $\mathbf{T}_p$ (and hence the need for a fixed choice of embedding of the dual of $\mathbb{Z}_p$ in $\mathbf{C}_p^\times$).

Theorem 7.5. *We have*

$$H_0(X_p) = L^1(\mathbf{T}_p) + \mathcal{G}(\mathrm{UD}(\mathbb{Z}_p)).$$

Proof Let E denote the space of all exponential polynomials on $\mathbb{Z}_p$ of the form $\sum_{i=1}^{r} q_i(z)\beta_i^{-z}$ where each $q_i(z) \in \mathbb{C}_p[z]$ and each $\beta_i \in \mathbb{U}_p \backslash \mathbb{T}_p$. Then clearly E is dense in $(\mathrm{UD}(\mathbb{Z}_p), V)$ (just consider functions of the form $z \longmapsto q(z)\beta^{-z}$ for a fixed $\beta \in \mathbb{U}_p\backslash\mathbb{T}_p$ and then use the fact that the polynomial functions are dense in $\mathrm{UD}(\mathbb{Z}_p)$; see the remark following 2.1).

Let $\mathcal{G}: \mathrm{UD}(\mathbb{Z}_p) \longrightarrow (C_0(X_p), V)$ denote the Gelfand transform which is isometric for valuation W and continuous for valuation V on $\mathrm{UD}(\mathbb{Z}_p)$ (since $W(f) \geqslant V(f) - 1$). Now $\mathcal{G}(E) \subseteq H_0(X_p) \subseteq C_0(X_p)$ and $\mathcal{G}(E)$ is dense in $H_0(X_p)$ by Theorem 6.5 and Definition 7.1. Hence $\mathcal{G}(\mathrm{UD}(\mathbb{Z}_p))$ is contained in and is dense in $H_0(X_p)$.

On the other hand let L denote the space of all locally constant functions on $\mathbb{Z}_p$ so that $L \subseteq \mathrm{UD}(\mathbb{Z}_p)$ and $\mathcal{G}(L)$ consists of the functions h on X_p which are zero on $\mathbf{V}_p \cup \{\infty\}$ and have finite support on $\mathbf{T}_p$ (clearly $h' = 0$). Now from the above $\mathcal{G}(L) \subseteq H_0(X_p)$ and clearly $\overline{\mathcal{G}(L)}$ (the closure of $\mathcal{G}(L)$ in $H_0(X_p)$) consists of the functions h on X_p which are zero on $\mathbf{V}_p \cup \{\infty\}$ and for which $\lim_\omega h(\omega) = 0$. Thus $\overline{\mathcal{G}(L)} \subseteq H_0(X_p)$ can be naturally identified with $L^1(\mathbf{T}_p)$ (see remark following 7.3). Thus $L^1(\mathbf{T}_p) + \mathcal{G}(\mathrm{UD}(\mathbb{Z}_p)) \subseteq H_0(X_p)$.

Now suppose that $h \in H_0(X_p)$. Then clearly $h|_{\mathbf{V}_p \cup \{\infty\}} \in H_0(\mathbf{V}_p)$ and so by 1.5(2) is of the form $-\theta(g)$ for some $g \in C(\mathbb{Z}_p)$. Choose $f \in \mathrm{UD}(\mathbb{Z}_p)$ with $f' = g$. Then clearly $k = h - \mathcal{G}(f)$ vanishes on $\mathbf{V}_p \cup \{\infty\}$ (as $\mathcal{G}(f)(\alpha) = -\theta(f')(\alpha)$ for $\alpha \in \mathbf{V}_p \cup \{\infty\}$). Now $k \in H_0(X_p)$ with $k(\alpha) = 0$ for all $\alpha \in \mathbf{V}_p \cup \{\infty\}$ and so by Lemma 7.3(2) $\lim_\omega v_p(k(\omega)) = \infty$. Hence $k \in \overline{\mathcal{G}(L)} \cong L^1(\mathbf{T}_p)$. Therefore $h = k + \mathcal{G}(f)$ belongs to $L^1(\mathbf{T}_p) + \mathcal{G}(\mathrm{UD}(\mathbb{Z}_p))$ as required. ◇◇

Remark 7.6. Let $J \in \mathrm{UD}(\mathbb{Z}_p)'$ be an integral with $J \neq I$ so that there exists a measure $\mu \in M(\mathbb{Z}_p)$ with $\mu \neq 0$ such that $J(f) = I(f) + \mu(f')$ for all $f \in \mathrm{UD}(\mathbb{Z}_p)$ (see remark following 2.7). In particular, for $\omega \in \mathbf{T}_p$, we have

$$J(z\varphi_\omega) = \hat{z}(\omega) + \mu(\varphi_\omega).$$

Now suppose that $\omega \longmapsto J(z\varphi_\omega)$ is the restriction to $\mathbf{T}_p$ of $h \in H_0(X_p)$ (say). Then by Theorem 7.5 we have that there exists $g \in \mathrm{UD}(\mathbb{Z}_p)$ and $k \in L^1(\mathbf{T}_p)$ with $\mu(\varphi_\omega) = \hat{g}(\omega) + k(\omega)$ for all $\omega \in \mathbf{T}_p$. Hence $\hat{g}(\omega) = \mu(\varphi_\omega) - k(\omega)$ has the asymptotic behaviour on $\mathbf{T}_p$ given by Lemma 1.14 (since $k(\omega) \longrightarrow 0$) and this contradicts the asymptotic behaviour given by Theorem 4.8 in [36]. (Note that this theorem extends, by continuity, from $\mathcal{G}(\mathrm{UD}(\mathbb{Z}_p))$ to $H_0(X_p)$.)

Therefore it is clear that I is the *unique* integral $J \in \mathrm{UD}(\mathbb{Z}_p)'$ for which $J(f\varphi_\omega) \longrightarrow 0$ 'weakly' for all $f \in \mathrm{UD}(\mathbb{Z}_p)$ (see remark following 7.3). Thus the H_0-Fourier theory appears naturally as the amalgam of the L^1-Fourier theory and the I-Fourier theory (note that $L^1(\mathbf{T}_p) \cap \mathcal{G}(\mathrm{UD}(\mathbb{Z}_p)) = \mathcal{G}(\mathrm{Int}\,(\mathbb{Z}_p))$ by Theorem 1).

Remark 7.7. (See [9, 10] and 5.3.) If $f \in \mathrm{UD}(\mathbb{Z}_p)$ is locally analytic then $\mathcal{G}(f)|_{\mathbf{V}_p \cup \{\infty\}} \in H_0(\mathbf{V}_p)$ extends analytically out of $\mathbf{V}_p \cup \{\infty\}$ to a larger disc around ∞ which thus includes all but finitely many $\omega \in \mathbf{T}_p$ and agrees with the values $\hat{f}(\omega)$ on all but finitely many of these (and vice versa). The remaining $\hat{f}(\omega)$ are not determined by this analytic continuation (indeed they are not determined by f' and $\mathcal{G}(f)(\alpha) = -\theta(f')(\alpha)$ for $\alpha \in \mathbf{V}_p \cup \{\infty\}$).

Theorem 7.8.

1. *We define the* Fourier transform $\mathcal{F}: (H_0(X_p), v) \longrightarrow (C(\mathbb{Z}_p), v)$ *by putting, for all* $g \in H_0(X_p)$,

$$\mathcal{F}(g) = \sum_{\omega} g(\omega)\varphi_{\omega^{-1}} \in C(\mathbb{Z}_p)$$

(a uniformly convergent 'series'). Clearly $v(\mathcal{F}(g)) \geqslant v(g)$ *and* $\mathcal{F}$ *is a continuous homomorphism of* $\mathbb{C}_p$*-Banach spaces.*

2. $\mathcal{F}$ *is surjective but not injective.*
3. $\ker(\mathcal{F})$ *consists of all* $g \in H_0(X_p)$ *such that there exists* $\{b_n\}$ *with* $v_p(b_n) + \log_p n \longrightarrow \infty$ *and* $g(\omega) = \sum_{n=0}^{\infty} b_n(\omega - 1)^n$ *for all* $\omega \in \mathbf{T}_p$.
4. $\mathcal{F}(\hat{f}) = f$ *for all* $f \in \mathrm{UD}(\mathbb{Z}_p)$ *(note that, by* 7.5, $\mathcal{G}(f) \in H_0(X_p)$ *and* $\mathcal{G}(f)|_{\mathbf{T}_p} = \hat{f}$*).*
5. $\mathcal{F}$ *induces an isometric isomorphism of* $\mathbb{C}_p$*-Banach spaces:*

$$\overline{\mathcal{F}}: (H_0(X_p)/\ker(\mathcal{F}), \bar{v}) \longrightarrow (C(\mathbb{Z}_p), v).$$

Proof This follows immediately from the L^1- and J-Fourier theories and the fact that $H_0(X_p) = L^1(\mathbf{T}_p) + \mathcal{G}(\mathrm{UD}(\mathbb{Z}_p))$ (Theorem 7.5; see also [13, 34]). ◊ ◊

Remark 7.9. In fact, as we will see from Theorem 7.10, it is possible to introduce multiplications $*, \ominus$ into $H_0(X_p)$ which correspond to the multiplications $\times, \circledast$ (respectively) on $C(\mathbb{Z}_p)$ under 7.8(5) above.

Theorem 7.10. *Let* $(\mathrm{UD}(\mathbb{Z}_p)_{\mathrm{spec}}, *, W)$ *denote the spectral completion of the* $\mathbb{C}_p$*-Banach algebra* $(\mathrm{UD}(\mathbb{Z}_p), *, V)$ *with respect to the valuation* W*. Then the Gelfand transform* $\mathcal{G}: (\mathrm{UD}(\mathbb{Z}_p), *, W) \longrightarrow (C_0(X_p), \times, v)$ *extends by continuity to an isometric isomorphism of* $\mathbb{C}_p$*-Banach algebras*

$$\mathcal{G}: (\mathrm{UD}(\mathbb{Z}_p)_{\mathrm{spec}}, *, W) \longrightarrow (H_0(X_p), \times, v).$$

Proof The Gelfand transform $\mathcal{G}$ is an isometric homomorphism of normed $\mathbb{C}_p$-algebras with image dense in the $\mathbb{C}_p$-Banach algebra $(H_0(X_p), \times, v) \subseteq (C_0(X_p), \times, v)$ (see the proof of Theorem 7.5). The result now follows. ◊ ◊

Remark 7.11. We have $\mathcal{G}(\overline{\mathrm{Int}(\mathbb{Z}_p)}) = L^1(\mathbf{T}_p)$ under the isomorphism $\mathcal{G}$ of Theorem 7.10. In fact we can easily characterize $\mathcal{G}(\mathrm{Int}(\mathbb{Z}_p))$ in $L^1(\mathbf{T}_p)$ (see [34]).

Remark 7.12. The multiplications $*$, $\circledast$, and $\times$ on $\mathrm{UD}(\mathbb{Z}_p)$ are all continuous with respect to W and so extend by continuity to $\mathrm{UD}(\mathbb{Z}_p)_{\mathrm{spec}}$ and hence can be transferred to $H_0(X_p)$ via $\mathcal{G}$ (of course $*$ on $\mathrm{UD}(\mathbb{Z}_p)$ corresponds to $\times$ on $H_0(X_p)$).

Now the identity map $(\mathrm{UD}(\mathbb{Z}_p), W) \longrightarrow (\mathrm{UD}(\mathbb{Z}_p), v)$ is continuous (since $v(f) \geqslant W(f)$ for all $f \in \mathrm{UD}(\mathbb{Z}_p)$) and so extends by continuity to a map of the corresponding completions

$$(H_0(X_p), v) \cong (\mathrm{UD}(\mathbb{Z}_p)_{\mathrm{spec}}, W) \twoheadrightarrow (C(\mathbb{Z}_p), v)$$

which can of course be naturally identified with the Fourier transform $\mathcal{F}$. We thus obtain a justification of the remark following Theorem 7.8.

Note also that differentiation $D\colon (\mathrm{UD}(\mathbb{Z}_p), W) \twoheadrightarrow (C(\mathbb{Z}_p), v)$ is continuous (since $v(f') \geqslant W(f)$ for all $f \in \mathrm{UD}(\mathbb{Z}_p)$) and so extends by continuity to a map

$$(H_0(X_p), v) \cong (\mathrm{UD}(\mathbb{Z}_p)_{\mathrm{spec}}, W) \twoheadrightarrow (C(\mathbb{Z}_p), v) \underset{\theta}{\cong} (H_0(\mathbf{V}_p), v)$$

(see 1.5) which can of course be naturally identified with 'minus the restriction map'. It is now easy to obtain 'explicit formulae' for the multiplications $*$ and $\ominus$ on $H_0(X_p)$ which correspond to $\times$ and $\circledast$ respectively on $\mathrm{UD}(\mathbb{Z}_p)$.

Remark 7.13. Note that $L^1(\mathbf{T}_p) \subseteq H_0(X_p)$ is clearly a (closed) ideal under $\times$ (the kernel of the restriction map mentioned in 7.12 above), is a subalgebra under $*$ (see 1.6) and is *not* closed under $\ominus$ (in fact

$$L^1(\mathbf{T}_p) \cap \mathcal{G}(\mathrm{UD}(\mathbb{Z}_p)) = \mathcal{G}(\mathrm{Int}(\mathbb{Z}_p))$$

and $1 \in \mathrm{Int}(\mathbb{Z}_p)$ while $1 \circledast 1 = z + 1 \notin \mathrm{Int}(\mathbb{Z}_p)$).

8 K-rationality

Finally, we briefly consider the 'action of Galois' in the situation described above (see also [37] for some further results and [14] for a related but rather different approach.) Let K be a closed subfield of $\mathbb{C}_p$, $\mathbb{Q}_p \subseteq K \subseteq \mathbb{C}_p$, and let N be the Galois group of $\mathbb{C}_p$ over K. We will denote the action of $\gamma \in N$ on $x \in \mathbb{C}_p$ by $x \longmapsto x^\gamma$. N acts on the various function spaces we have considered in a natural way: if 'f is a function with variable x' then we define $f^\gamma(x) = f(x^{\gamma^{-1}})^\gamma$ (note the simplification which occurs when $x \in \mathbb{Z}_p$) for all $\gamma \in N$. If F is a 'function space' then we denote the set of all N-invariant elements of F by F^N. Clearly the various Fourier transforms etc. considered above commute with the action of N.

Theorem 8.1. *The Fourier transform* $\mathcal{F}: (L^1(\mathbf{T}_p), *, v) \longrightarrow (C(\mathbb{Z}_p), \times, v)$ *induces a continuous homomorphism of K-Banach algebras*

$$\mathcal{F}^N: (L^1(\mathbf{T}_p)^N, *, v) \longrightarrow (C(\mathbb{Z}_p)^N = C(\mathbb{Z}_p, K), \times, v)$$

(where $C(\mathbb{Z}_p, K)$ denotes the K-valued continuous functions on $\mathbb{Z}_p$) by restriction.

If further K has 'non-zero different over $\mathbb{Q}_p$' (see [14, 36]) then the spectral valuation W is equivalent to V on $\mathrm{Int}(\mathbb{Z}_p, \mathrm{K})$ *and $\mathcal{F}^N$ induces an isometric isomorphism of K-Banach algebras from $(L^1(\mathbf{T}_p)^N, *, v)$ onto* $(\mathrm{Int}(\mathbb{Z}_p, \mathrm{K}), \times, \mathrm{W})$ *(such that the multiplications $\times$ on $L^1(\mathbf{T}_p)^N$ and $*$ on* $\mathrm{Int}(\mathbb{Z}_p, \mathrm{K})$ *also correspond).*

Proof This will follow directly from the proof of Theorem 8.3 (or see [36]). ◊ ◊

Remark 8.2. If K has 'zero different over $\mathbb{Q}_p$' then the situation is radically different. For example, at least if $K \supseteq \mathbb{Q}_p(\mathbf{T}_p)$, then $\mathcal{F}^N$ is surjective but not injective and we obtain the obvious analogue of Theorem 1.8 for $\mathcal{F}^N$ (see [14]).

Theorem 8.3.

1. *The isometric isomorphism of $\mathbb{C}_p$-Banach algebras (see Theorem 7.10)*

$$\mathcal{G}: (\mathrm{UD}(\mathbb{Z}_p)_{\mathsf{spec}}, *, W) \longrightarrow (H_0(X_p), \times, v)$$

induces, by restriction, an isometric isomorphism of K-Banach algebras

$$\mathcal{G}^N: (\mathrm{UD}(\mathbb{Z}_p, K)_{\mathsf{spec}} = \mathrm{UD}(\mathbb{Z}_p)^N_{\mathsf{spec}}, *, W) \longrightarrow (H_0(X_p)^N, \times, v)$$

(where of course the multiplications $\circledast$, $\times$ on $\mathrm{UD}(\mathbb{Z}_p, K)_{\mathsf{spec}}$ and $\ominus$, $*$ respectively on $H_0(X_p)^N$ also correspond).

2. $\mathrm{UD}(\mathbb{Z}_p, K)_{\mathsf{spec}} = \mathrm{UD}(\mathbb{Z}_p, K) \Longleftrightarrow$ *K has non-zero different over $\mathbb{Q}_p$.*

Proof

1. The required isomorphism is obtained by restricting $\mathcal{G}$ to the corresponding N-invariant elements. It remains to show that $\mathrm{UD}(\mathbb{Z}_p, K) = \mathrm{UD}(\mathbb{Z}_p)^N$ is dense in the K-Banach algebra $(\mathrm{UD}(\mathbb{Z}_p)^N_{\mathrm{spec}}, *, W)$ and so its spectral completion $\mathrm{UD}(\mathbb{Z}_p, K)_{\mathrm{spec}}$ can be naturally identified with $\mathrm{UD}(\mathbb{Z}_p)^N_{\mathrm{spec}}$. By applying the Gelfand transform $\mathcal{G}$ it is therefore enough to show that $\mathcal{G}(\mathrm{UD}(\mathbb{Z}_p, K))$ is dense in $H_0(X_p)^N$. Let $h \in H_0(X_p)^N$. Then $h|_{\mathbf{V}_p \cup \{\infty\}} = -\theta(g) \in H_0(\mathbf{V}_p)$ for some $g \in C(\mathbb{Z}_p)$. Since θ clearly commutes with the action of N (see 1.5) we have $g \in C(\mathbb{Z}_p)^N = C(\mathbb{Z}_p, K)$. Now choose $f \in \mathrm{UD}(\mathbb{Z}_p, K)$ with $f' = g$. Then $k = h - \mathcal{G}(f)$ vanishes on $\mathbf{V}_p \cup \{\infty\}$ and therefore clearly $k \in L^1(\mathbf{T}_p)^N$. Let S denote the space of all K-valued locally constant functions on $\mathbb{Z}_p$ so that $S \subseteq \mathrm{UD}(\mathbb{Z}_p, K)$ and clearly $\mathcal{G}(S)$ is dense in $L^1(\mathbf{T}_p)^N$. The result now follows easily.
2. See [36].

This completes the proof of Theorem 8.3. ◊ ◊

Remark 8.4. Thus in the case when K has non-zero different over $\mathbb{Q}_p$ the 'Fourier transform' provides a one to one correspondence between substantial spaces of *functions* on $\mathbf{T}_p$ and $\mathbb{Z}_p$.

Remark 8.5. Let R denote the space of all rational functions on X_p with coefficients in K vanishing at ∞ and with no pole in X_p. Then clearly $R \subseteq H_0(X_p)^N$ and in fact R is dense in $H_0(X_p)^N$. For, by Theorem 6.5, $R \subseteq \mathcal{G}(\mathrm{UD}(\mathbb{Z}_p))$ and further by Theorem 7.8(4), $R \subseteq \mathcal{G}(\mathrm{UD}(\mathbb{Z}_p, K))$ (clearly for $r \in R$, $\mathcal{F}(r)$ takes values in K). Finally R is dense in $\mathcal{G}(\mathrm{UD}(\mathbb{Z}_p, K))$ and so in $H_0(X_p)^N$ (by Theorem 8.3; $K[z] \cdot (1+p)^z$ is V-dense and so W-dense in $\mathrm{UD}(\mathbb{Z}_p, K)$ and has Gelfand transform contained in R).

References

1. Y. Amice, 'Dual d'un espace $H(D)$ et transformation de Fourier', Groupe de travail Analyse Ultramétrique, Paris (1974), No. 7.

2. Y. Amice, 'Les Nombres p-Adiques', Presses Universitaires de France, (1975).

3. Y. Amice, 'Duals', in 'Proceedings of the conference on p-adic analysis', University of Nijmegen, (1978)', 1–15.

4. Y. Amice and A. Escassut, 'Sur la non-injectivité de la transformation de Fourier p-adiqué relative à $\mathbb{Z}_p$', CR Acad. Sci. Paris **278**, Série A (1974), 583–585.

5. D. Barsky, 'Fonctions k-lipschitziennes sur un anneau local et polynômes à valuers entieres', Bull. Soc. Math. France **101** (1973), 397–411.

6. G. Borm, 'p-adic Fourier theory', (thesis), University of Nijmegen (1988).

7. F. Bruhat, 'Intégration p-adique', Séminaire Bourbaki 14e année (1961/62), No. **229**.

8. J. W. S. Cassels, 'Local Fields', Cambridge University Press (1986).

9. P. Cassou-Noguès, 'Formes linéaires p-adiques et prolongement analytique', (thesis), University of Bordeaux (1971).

10. P. Cassou-Noguès, 'Formes linéaires p-adiques et prolongement analytique', CR Acad. Sci. Paris **274** Sér. A (1972), 5–8.

11. J. Dieudonné, 'Sur les functions continues p-adiques', Bull. Sci. Math. **68** (1944), 79–95.

12. A. Escassut, 'Propriétés spectrales en analyse non-archimédienne', CR Acad. Sci. Paris **278** Sér. A (1974), 1387–1389.

13. J. Fresnel and B. de Mathan, 'Sur la transformation de Fourier p-adique', Séminaire de Théorie des Nombres, Bordeaux (1972–73), No. **21**.

14. J. Fresnel and B. de Mathan, 'Algèbres L^1 p-adiques', Bull. Soc. Math. France, **106** (1978), 225–260.

15. J. Fresnel and M. Matignon, 'Produit tensoriel topologique de corps values', Can. J. Math. **XXXV** (1983), 218–273.

16. A. M. M. Gommers, 'Non-archimedean harmonic analysis on groups without Haar measure', (thesis), University of Nijmegen (1979).

17. J. Hily, 'Interpolation p-adique', Séminaire Delange-Pisot (Théorie des nombres) (1962/3), No. **15**.

18. N. Koblitz, 'p-adic Numbers, p-adic Analysis and Zeta Functions', Springer-Verlag, Berlin (1977).

19. N. Koblitz, 'p-adic Analysis: a short course on recent work', Cambridge University Press (1980).

20. T. Kubota and H. W. Leopoldt, 'Eine p-adische theorie der Zetawerte (I)', J. reine angew. Math. **214/5** (1964), 328–339.

21. L. H. Loomis, 'An Introduction to Abstract Harmonic Analysis', Van Nostrand, New York (1953).

22. K. Mahler, 'An interpolation series for continuous functions of a p-adic variable', J. reine angew. Math. **199** (1958), 23–34, and **208** (1961), 70–72 (correction).

23. K. Mahler, 'p-adic Numbers and their Functions', Cambridge University Press (1981).

24. M. van der Put, 'Difference equations over p-adic fields', Math. Ann. **198** (1972), 189–203.

25. A. C. M. van Rooij, 'Non-Archimedean Functional Analysis', Dekker, New York (1978).

26. W. H. Schikhof, 'Non-archimedean harmonic analysis', (thesis), University of Nijmegen (1967).

27. W. H. Schikhof, 'Ultrametric Calculus', Cambridge University Press (1984).

28. K. Shiratani and S. Yokoyama, 'An application of p-adic convolutions', Mem. Fac. Sci. Kyushu Univ. Ser. A. **36** (1982), 73–83.

29. M. H. Taibleson, 'Fourier analysis on local fields', Mathematical Notes **15**, Princeton University Press (1975).

30. F. Tomàs, 'Integracion p-adica', Boletin de la Sociedad Matematica Mexicana (1962), 1–38.

31. A. Volkenborn, 'Ein p-adisches Integral und seine Anwendungen I,II', Manuscripta Math. **7** (1972), 341–373, and **12** (1974), 17–46.

32. C.S. Weisman, 'On p-adic differentiability', J. Number Theory **9** (1977), 79–86.

33. C. F. Woodcock, 'p-adic Fourier analysis', (thesis), University of Oxford (1971).

34. C. F. Woodcock, 'Parseval's formula and p-adic interpolation', in 'Proceedings of the conference on p-adic analysis', University of Nijmegen (1978), 205–217.

35. C. F. Woodcock, 'Convolutions on the ring of p-adic integers', J. London Math. Soc. (2) **20** (1979), 101–108.

36. C. F. Woodcock, 'A spectral valuation on the ring of p-adic integers', J. London Math. Soc. (2) **25** (1982), 223–234.

37. C. F. Woodcock, 'Spectral properties of tame functions on the ring of p-adic integers', J. London Math. Soc. (2) **30** (1984), 407–418.

38. C. F. Woodcock, 'A two variable Riemann zeta function', J. Number Theory **27** (1987), 212–221.

The manufacturer's authorised representative in the EU for product safety is Oxford University Press España S.A. of el Parque Empresarial San Fernando de Henares, Avenida de Castilla, 2 – 28830 Madrid (www.oup.es/en or product.safety@oup.com). OUP España S.A. also acts as importer into Spain of products made by the manufacturer.

www.ingramcontent.com/pod-product-compliance
Ingram Content Group UK Ltd.
Pitfield, Milton Keynes, MK11 3LW, UK
UKHW022153230426
12049UKWH00003BA/82
9780198535942